The ESSENTIALS® of

Organic Chemistry II

Staff of Research and Education Association
Dr. M. Fogiel, Director

This book is a continuation of *"THE ESSENTIALS OF ORGANIC CHEMISTRY I"* and begins with Chapter 14. It covers the usual course outline of Organic Chemistry II. Earlier basic topics are covered in *"THE ESSENTIALS OF ORGANIC CHEMISTRY I."*

Research & Education Association
61 Ethel Road West
Piscataway, New Jersey 08854

THE ESSENTIALS®
OF ORGANIC CHEMISTRY II

Printed in the United States of America

Library of Congress Control Number 00-109276

International Standard Book Number 0-87891-617-2

IV-1

WHAT "THE ESSENTIALS" WILL DO FOR YOU

This book is a review and study guide. It is comprehensive and it is concise.

It helps in preparing for exams and in doing homework, and remains a handy reference source at all times.

It condenses the vast amount of detail characteristic of the subject matter and summarizes the **essentials** of the field.

It will thus save hours of study and preparation time.

The book provides quick access to the important facts, principles, theorems, concepts, and equations in the field.

Materials needed for exams can be reviewed in summary form – eliminating the need to read and re-read many pages of textbook and class notes. The summaries will even tend to bring detail to mind that had been previously read or noted.

This "ESSENTIALS" book has been prepared by experts in the field, and has been carefully reviewed to ensure its accuracy and maximum usefulness.

Dr. Max Fogiel
Program Director

CONTENTS

This book is a continuation of *THE ESSENTIALS OF ORGANIC CHEMISTRY I* and begins with Chapter 14. It covers the usual course outline of Organic Chemistry II. Earlier/basic topics are covered in *THE ESSENTIALS OF ORGANIC CHEMISTRY I.*

THE PERIODIC TABLE

METALS — NONMETALS

KEY

	4 IVA IVB	
Atomic Number →	22	← Group Classification
	Ti	← Symbol
Atomic Weight → ( ) indicates most stable or best known isotope	47.88	

TRANSITIONAL METALS (groups 3–11)

1 IA IA	2 IIA IIA	3 IIIA IIIB	4 IVA IVB	5 VA VB	6 VIA VIB	7 VIIA VIIB	8 VIIIA VIII	9 VIIIA VIII	10 VIIIA VIII	11 IB IB	12 IIB IIB	13 IIIB IIIA	14 IVB IVA	15 VB VA	16 VIB VIA	17 VIIB VIIA	18 VIII 0
1 H 1.008																	2 He 4.003
3 Li 6.941	4 Be 9.012											5 B 10.811	6 C 12.011	7 N 14.007	8 O 15.999	9 F 18.998	10 Ne 20.180
11 Na 22.990	12 Mg 24.305											13 Al 26.982	14 Si 28.086	15 P 30.974	16 S 32.066	17 Cl 35.453	18 Ar 39.948
19 K 39.098	20 Ca 40.078	21 Sc 44.956	22 Ti 47.88	23 V 50.942	24 Cr 51.996	25 Mn 54.938	26 Fe 55.847	27 Co 58.933	28 Ni 58.693	29 Cu 63.546	30 Zn 65.39	31 Ga 69.723	32 Ge 72.61	33 As 74.922	34 Se 78.96	35 Br 79.904	36 Kr 83.8
37 Rb 85.468	38 Sr 87.62	39 Y 88.906	40 Zr 91.224	41 Nb 92.906	42 Mo 95.94	43 Tc (97.907)	44 Ru 101.07	45 Rh 102.906	46 Pd 106.4	47 Ag 107.868	48 Cd 112.411	49 In 114.818	50 Sn 118.710	51 Sb 121.757	52 Te 127.60	53 I 126.905	54 Xe 131.29
55 Cs 132.905	56 Ba 137.327	57 La 138.906	72 Hf 178.49	73 Ta 180.948	74 W 183.84	75 Re 186.207	76 Os 190.23	77 Ir 192.22	78 Pt 195.08	79 Au 196.967	80 Hg 200.59	81 Tl 204.383	82 Pb 207.2	83 Bi 208.980	84 Po (208.982)	85 At (209.982)	86 Rn (222.018)
87 Fr (223.020)	88 Ra (226.025)	89 Ac (227.028)	104 Unq (261.11)	105 Unp (262.114)	106 Unh (263.118)	107 Uns (262.12)	108 Uno (265)	109 Une (266)	110 Uun (269)								
Alkali Metals	Alkaline Earth Metals															Halo-gens	Noble Gases

LANTHANIDE SERIES	58 Ce 140.115	59 Pr 140.908	60 Nd 144.24	61 Pm (144.913)	62 Sm 150.36	63 Eu 151.965	64 Gd 157.25	65 Tb 158.925	66 Dy 162.50	67 Ho 164.930	68 Er 167.26	69 Tm 168.934	70 Yb 173.04	71 Lu 174.967
ACTINIDE SERIES	90 Th 232.038	91 Pa 231.036	92 U 238.029	93 Np (237.048)	94 Pu (244.064)	95 Am (243.061)	96 Cm (247.070)	97 Bk (247.070)	98 Cf (251.080)	99 Es (252.083)	100 Fm (257.095)	101 Md (258.1)	102 No (259.101)	103 Lr (262.11)

CHAPTER 14

ARENES

14.1 STRUCTURE AND NOMENCLATURE

Arenes are compounds that contain both aromatic and aliphatic units.

The simplest of the alkyl benzenes, methyl benzene, has the common name toluene. Compounds that have longer side chains are named by adding the word "benzene" to the name of the alkyl group.

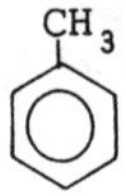

CH_3 Toluene

$CH_2CH(CH_3)CH_3$ Isobutylbenzene

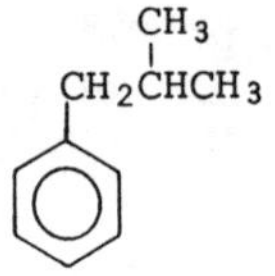

C_2H_5, $CH(CH_3)CH_3$ m-Ethylisopropylbenzene

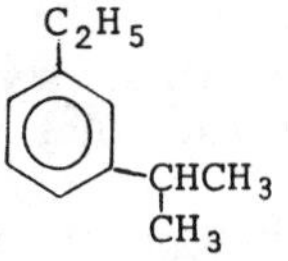

The simplest of the dialkyl benzenes, the dimethyl benzenes, have the common name xylenes. Dialkyl benzenes that contain one methyl group are named as derivatives of toluene.

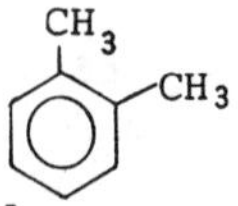

o-Xylene

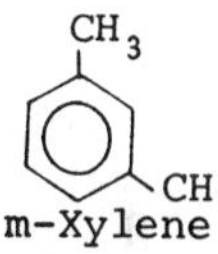

m-Xylene

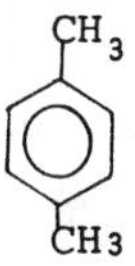

p-Xylene

p-Ethyltoluene (CH_3, CH_2CH_3)

A compound that contains a complex side chain is named as a phenyl alkane (C_6H_5 = phenyl). Compounds that contain more than one benzene ring are named as derivatives of alkanes.

$CH_3CH(CH_3)CH(C_6H_5)CH_2CH_3$

2-Methyl-3-phenylpentane

C_6H_5-CH_2CH_2-C_6H_5

1,2-Diphenylethane

Styrene is the name given to the simplest alkenyl benzene. Others are named as substituted alkenes. Alkynyl benzenes are named as substituted alkynes.

C_6H_5-CH=CH_2

Styrene
(Vinylbenzene)
(Phenylethylene)

C_6H_5-CH_2-CH=CH_2

Allyl benzene
(3-Phenylpropene)

C_6H_5-CH=CH

Phenylacetylene

14.2 PHYSICAL PROPERTIES OF ARENES

Alkyl benzenes are insoluble in water, but they are soluble in non-polar solvents like ether; they are generally less dense than water, and their boiling points rise with increasing molecular weight.

Melting points not only depend on the molecular weight, but also on the molecular shape. There exist relationships between melting points and structures of aromatic compounds; for example, in isomeric disubstituted benzenes, the para isomer generally melts at a considerably higher temperature than do the other two. Since dissolution, like melting, involves overcoming the intermolecular forces of the crystal, the para isomer which forms the most stable crystals, is the least soluble in a given solvent.

The higher melting point and lower solubility of a para isomer is an example of the effect of molecular symmetry on intracrystalline forces. The more symmetrical a compound is, the better it fits into a crystal lattice, hence the melting point is increased, and the solubility is lowered.

14.3 PREPARATION OF ALKYLBENZENES

A) Attachment of alkyl groups: Friedel-Crafts alkylation

$$C_6H_6 + RX \xrightarrow{\text{Lewis acid}} C_6H_5R + HX \quad \text{R may rearrange}$$

Lewis acid: $AlCl_3$, BF_3, HF, etc.
Ar-X cannot be used in place of R-X

Mechanism of Friedel-Crafts alkylation.

a) $RCl + AlCl_3 \rightleftharpoons AlCl_4^- + R^{\oplus}$ Carbonium ions from alkyl halides

b) $R^{\oplus} + C_6H_6 \rightleftharpoons C_6H_5^{\oplus}\langle^{R}_{H}$

c) $C_6H_5^{\oplus}\langle^{R}_{H} + AlCl_4^- \rightleftharpoons C_6H_5R + HCl + AlCl_3$

A carbonium ion may: (1) eliminate a hydrogen ion to form an alkene; (2) rearrange to form a more stable carbonium ion; (3) combine with a basic molecule or a negative ion; (4) form a larger carbonium ion by adding to an alkene; (5) abstract a hydride ion from an alkene; (6) alkylate an aromatic ring.

The use of the Friedel-Crafts alkylation is limited by: (1) the risk of polysubstitution; (2) a possibility of rearrangement of the alkyl group; (3) the impossibility of replacement of the alkyl halides by the aryl halides; (4) aromatic rings that are less reactive than halobenzenes do not undergo Friedel-Crafts alkylation; (5) aromatic rings that contain $-NH_2$, -NHR, or $-NR_2$ groups do not undergo Friedel-Crafts alkylation.

B) Conversion of Side Chain

a)

$$C_6H_5\text{-}\overset{O}{\overset{\|}{C}}\text{-}R \xrightarrow[\text{or } N_2H_4\text{, base, heat}]{\text{Zn(Hg), HCl, heat}} C_6H_5CH_2R$$

A Ketone

Clemmensen or Wolff-Kishner reduction.

b)

$$C_6H_5-CH{=}CHR \xrightarrow{H_2/Ni} C_6H_5-CH_2CH_2R$$

C) **Dehydrocyclization Reaction**

$$\underset{\text{Heptane}}{CH_3(CH_2)_5CH_3} \xrightarrow{Cr_2O_3;550°C} \underset{\text{Toluene}}{C_6H_5-CH_3} + 4H_2$$

D) **Ullmann Reaction**

$$2R-C_6H_4-I \xrightarrow[\text{Heat}]{Cu} R-C_6H_4-C_6H_4-R + CuI_2$$

E) **Electrophilic Aromatic Substitution**

$$\underset{\text{Benzene}}{C_6H_6} + \underset{\text{Ethyl alcohol}}{CH_3CH_2OH} \xrightarrow{H_2SO_4,\text{heat}} \underset{\text{Ethylbenzene}}{C_6H_5-CH_2CH_3} + H_2O$$

F) **Hydrogenation**

$$\underset{\text{Styrene}}{C_6H_5-CH{=}CH_2} + H_2 \xrightarrow{Pt,\text{heat},\text{pressure}} \underset{\text{Ethylbenzene}}{C_6H_5-CH_2CH_3}$$

14.4 REACTIONS OF ALKYLBENZENES

A) **Hydrogenation**

$$\underset{\text{Ethylbenzene}}{C_6H_5-CH_2CH_3} + 3H_2 \xrightarrow{Ni,Pt \text{ or } Pd} \underset{\text{Ethylcyclohexane}}{C_6H_{11}-CH_2CH_3}$$

B) **Oxidation**

$$\underset{\text{Ethylbenzene}}{C_6H_5-CH_2CH_3} \xrightarrow[\text{(or } K_2Cr_2O_7, \text{ or dil.} HNO_3)]{KMnO_4} \underset{\text{Benzoic acid}}{C_6H_5-COOH} \quad (+\ CO_2)$$

The oxidation reaction is used for two purposes: (a) synthesis of carboxylic acids, and (b) identification of alkylbenzenes.

C) **Substitution in the ring. Electrophilic aromatic substitution.**

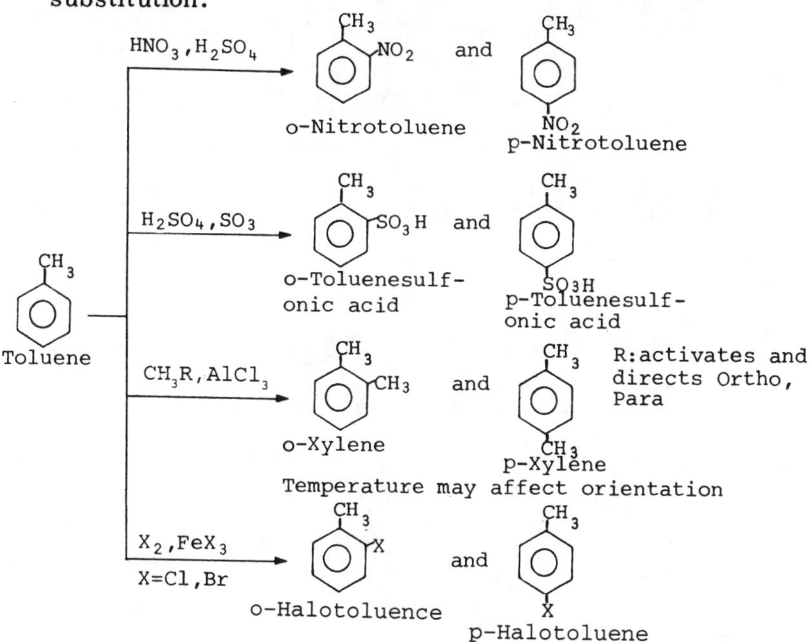

D) **Substitution in the side chain. Free-radical halogenation.**

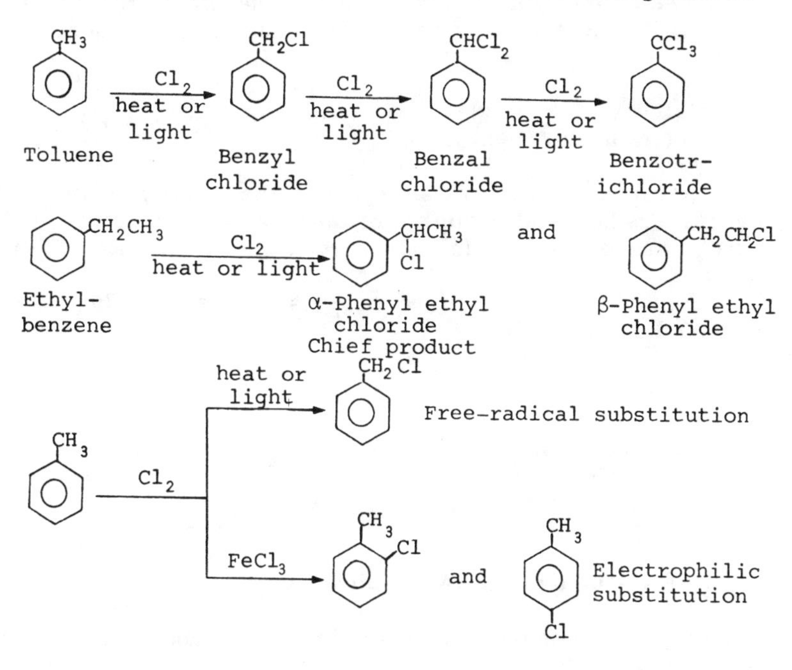

Hydrogen atoms attached to a carbon joined directly to an aromatic ring are called benzylic hydrogens.

$C_6H_5-\overset{|}{\underset{H}{C}}-$ Benzylic hydrogen

Ease of abstraction of hydrogen atoms: allylic, benzylic $> 3° > 2° > 1° > CH_4$, vinylic.

14.5 RESONANCE STABILIZATION OF BENZYL RADICALS

$$C_6H_5-CH_3 \longrightarrow C_6H_5-CH_2\cdot + H\cdot$$

Toluene → Benzyl radical

Ease of formation of free radicals: allyl, benzyl $> 3° > 2° > 1° > CH_3$, vinyl

The more stable the radical, the more rapidly it is formed; the less energy the radical contains, the more stable it is.

Stability of free radicals: allyl, benzyl $> 3° > 2° > 1° > CH_3$, vinyl

Resonance stabilizes and lowers the energy content of a benzyl radical. Through resonance the benzyl radical becomes more stable than the hydrocarbon it is formed from.

Benzyl and allyl free radicals are extremely reactive and unstable particles.

14.6 PREPARATION AND REACTIONS OF ALKENYLBENZENES

PREPARATION OF ALKENYLBENZENES

Alkenylbenzenes are aromatic hydrocarbons with a side

chain containing a double bond.

A) Dehydrogenation

$$C_6H_6 + CH_2{=}CH_2 \xrightarrow{H_3PO_4} C_6H_5CH_2CH_3 \xrightarrow[90\%\ yield]{Cr_2O_3 \cdot Al_2O_3, 600^\circ} C_6H_5CH{=}CH_2 + H_2$$

Ethylbenzene Styrene

B) Dehydrohalogenation and dehydration

$$C_6H_5CH(Cl)CH_3 \xrightarrow{KOH(alc), heat} C_6H_5CH{=}CH_2$$

1-Phenyl-1-chloroethane Styrene

Dehydrohalogenation

$$C_6H_5CH(OH)CH_3 \xrightarrow{ZnCl_2, heat} C_6H_5CH{=}CH_2$$

1-Phenylethanol Styrene

Dehydration

$$C_6H_5CH_2CH(Cl)CH_3 \xrightarrow{KOH (alc), heat} C_6H_5CH{=}CH{-}CH_3 + C_6H_5CH_2CH{=}CH_2 \xleftarrow{acid, heat} C_6H_5CH_2CH(OH)CH_3$$

1-Phenyl-2-chloropropane 1-Phenylpropene 3-Phenylpropene 1-Phenyl-2-propanol

$$C_6H_5CH_2{-}CH{=}CH_2 \xrightarrow{KOH, heat} C_6H_5CH{=}CH{-}CH_3$$

3-Phenylpropene (Allyl benzene) 1-Phenylpropene

A double bond is conjugated with the ring when it is separated from the benzene ring by one single bond. Conjugation gives unusual stability to a molecule. This stability has an effect on the orientation and the ease of elimination.

$C_6H_5{-}\overset{|}{C}{=}\overset{|}{C}{-}$ Double bond conjugated with ring.

REACTIONS OF ALKENYLBENZENES

The double bond shows higher reactivity than the resonance-stabilized benzene ring.

A) Substitution in the Ring

a) Catalytic hydrogenation

$$C_6H_5CH{=}CH_2 \xrightarrow[75\ \text{minutes}]{H_2/Ni,\,20°C,\,2\text{-}3\ \text{atm.}} C_6H_5CH_2CH_3 \xrightarrow[100\ \text{minutes}]{H_2/Ni,\,125°C,\,110\ \text{atm.}} C_6H_{11}CH_2CH_3$$

Styrene — Ethyl-benzene — Ethyl-cyclohexane

b) Ring halogenation

$$C_6H_5C_2H_5 \xrightarrow{Cl_2,\,FeCl_3} p\text{-}ClC_6H_4C_2H_5 \xrightarrow{Cl_2,\,heat} p\text{-}ClC_6H_4CHClCH_3 \xrightarrow{KOH} p\text{-}ClC_6H_4CH{=}CH_2$$

Ethylbenzene — p-Chlorostyrene

B) Addition to conjugated alkenylbenzenes: Orientation. Stability of benzylcation. In either electrophilic or free-radical additions, the first step takes place in the way that yields the more stable particle.

No peroxides $C_6H_5CH = CHCH_3 \xrightarrow{HBr} C_6H_5\overset{\oplus}{C}HCH_2CH_3 \xrightarrow{Br^{\ominus}}$

a benzylcation

$C_6H_5CH(Br)CH_2CH_3$

Peroxides present $C_6H_5CH = CHCH_3 \xrightarrow{Br^\bullet} C_6H_5\overset{\bullet}{C}H\underset{Br}{\underset{|}{C}}HCH_3 \xrightarrow{HBr}$

a benzyl free-radical

$C_6H_5CH_2\underset{Br}{\underset{|}{C}}HCH_3$

Stability of carbonium ions: 3° (benzyl) > 2° (allyl) > 1° > CH_3^+

Addition to conjugated alkenylbenzenes: reactivity.

Conjugated alkenylbenzenes are more stable than simple alkenes.

Conjugated alkenylbenzenes are much more reactive than simple alkenes toward both ionic and free-radical addition.

CHAPTER 15

ALDEHYDES AND KETONES

Carboxylic acids, aldehydes and ketones have a carboxylic group, $>C=O$ in common. The general formula for aldehydes is RCHO, and that for ketones is RCOR.

15.1 NOMENCLATURE (IUPAC SYSTEM)

A) Aldehydes: The longest continuous chain containing the carbonyl group is considered the parent structure and the "-e" ending of the corresponding alkane is replaced by "-al."

B) Ketones: The "-e" ending of the corresponding alkane is replaced by "-one."

Ex.

H | H–C=O	H | $CH_3C=O$	CH_3 | $CH_3-C=O$	CH_3 | $CH_3-CH_2-C=O$
methanal (formaldehyde)	ethanal (acetaldehyde)	propanone (dimethyl ketone)	butanone (methyl ethyl ketone)

15.2 PHYSICAL PROPERTIES

A) Formaldehyde and acetaldehyde are colorless gases.

B) The boiling points of aldehydes are much lower than those of corresponding alcohols, and their solubility in water decreases with an increase in carbon content.

C) Aldehydes are easily oxidized.

D) Acetone is a colorless liquid and although less dense than water, it is completely miscible with it.

E) Ketones and aldehydes cannot form intermolecular hydrogen bonding, and thus they have lower boiling points than alcohols and carboxilic acids of comparable molecular weight.

With an increase in the molecular size of the carbonyl compound, the influence of a non-polar alkyl group predominates, and solubility decreases.

15.3 PREPARATION OF ALDEHYDES

By removal of water from a primary alcohol group linked to the desired radical through:

A) Oxidation

$R-CH_2-OH$ + Mild oxidation $\rightarrow R-CHO + H_2O$

Ex.

$3CH_3-CH_2-OH + K_2Cr_2O_7/4H_2SO_4$, aq $\rightarrow 3CH_2-CHO$
ethanol ethanal
$+ Cr_2(SO_4)_3 + K_2SO_4 + 7H_2O$

B) Catalytic dehydrogenation

$R-CH_2-OH$ + Cu, 300°C $\rightarrow R-CHO + H_2$

By heating calcium salts of fatty acids containing the desired radical with calcium formate:

$(R-CO-O)_2Ca + (H-CO-O)_2Ca$, fuse $\rightarrow 2R-CHO + 2CaCO_3$

Ex. $CH_3-CO-O-Ca-O-OC-CH_3$
$+ H-CO-O-Ca-OOC-H \rightarrow 2CH_3-CHO + 2CaCO_3$
ethanal

By passing the vapors of the acid mixed with formic acid vapors over manganous oxide at 300°C.

$$R{-}CO{-}OH + H{-}CO{-}OH/(MnO, 300°C) \rightarrow R{-}CHO + CO_2 + H_2O$$

By the hydrolysis of the corresponding dihaloalkanes.

$$R{-}CHCl_2 + H_2O/(PbO, boil) \rightarrow R{-}CHO + 2HCl$$

By reaction of Grignard reagent with ethyl formate (ethyl orthoformate) or HCN in ether, followed by hydrolysis.

a) $R{-}MgX + H{-}CO{-}O{-}CH_2{-}CH_3 \rightarrow R{-}CHO + CH_3{-}CH_2{-}O{-}MgX$

b) $R{-}MgX + HCN \xrightarrow{ether} R{-}C(=N{-}Mg{-}X)H$

$R{-}C(=N{-}Mg{-}X)H + H_2O/2HX \rightarrow RCHO + MgX_2 + NH_4X$

By cleavage; glycols react with lead tetracetate in anhydrous benzene solution.

$$\underset{\displaystyle OH\quad OH}{R{-}CH{-}CH{-}R'} + Pb(CH_3COO)_4 \xrightarrow[anhydrous]{C_6H_6} R{-}CHO + R'{-}CHO + Pb(CH_3COO)_2 + 2CH_3COOH$$

15.4 PREPARATION OF KETONES

Removal of two hydrogen atoms from a secondary alcohol group linked to the desired radicals through:

A) Oxidation

RCH_2OH + mild oxidation $\rightarrow R \cdot CO \cdot R + H_2O$

Ex.

$$3(CH_3 \cdot CHOH \cdot CH_3) + Cr_2O_7 + 8H^+ \rightarrow 3(CH_3 \cdot CO \cdot CH_3) + 2Cr^{++} + 7H_2O$$

B) Catalytic dehydrogenation

$RCH + (Cu, 300°C) \rightarrow RCOR + H_2$

By reaction of Grignard reagent with esters (other than those of formic acid or acyl halides) or alkyl nitriles (or amides) in ether, followed by hydrolysis.

a) $R \cdot Mg \cdot X + R \cdot CO \cdot O \cdot Et$, ether $\rightarrow R \cdot CO \cdot R + Et \cdot O \cdot Mg \cdot X$

b) $R \cdot Mg \cdot X + R \cdot CN \rightarrow R{-}C({=}N{-}Mg{-}X)R$

$R{-}C({=}N{-}Mg{-}X)R + H_2O/2HX \rightarrow R{-}CO{-}R + MgX_2 + NH_4X$

By the hydrolysis of the corresponding dihaloalkanes

$R_2CCl_2 + H_2O/(PbO, boil) \rightarrow R_2C{=}O + 2HCl$

Ketones from carboxylic acids and their derivatives.

$$2R{-}COOH \xrightarrow[300°C]{MnO} R_2C = O + CO_2 + H_2O$$

Nucleophilic substitution of the halides by the alkyl or aryl group of organocadmium compounds.

$$2R{-}\underset{\underset{O}{\|}}{C}{-}X + R'{-}\underset{\underset{R'}{|}}{Cd} \rightarrow 2R{-}\underset{\underset{O}{\|}}{C}{-}R' + CdX_2$$

R = alkyl, aryl
R' must be aryl or 1° alkyl

Friedel-Crafts acylation of aromatic compounds

$$ArH + R{-}C(=O)Cl \xrightarrow[\text{or other Lewis acid}]{AlCl_3} Ar{-}\underset{\underset{O}{\|}}{C}{-}R + HCl$$

Cleavage of glycols having the hydroxyl group attached to tertiary carbon atoms.

$$R-\underset{OH}{\overset{R}{C}}-\underset{OH}{\overset{R'}{C}}-R' + Pb(CH_3COO)_4 \xrightarrow[\text{anhydrous}]{C_6H_6} R-\underset{O}{\overset{R}{\underset{\|}{C}}} + \overset{R'}{C}(=O)-R'$$

$$+ Pb(CH_3COO)_2 + 2CH_3COOH$$

15.5 REACTIONS OF ALDEHYDES AND KETONES

By oxidation when treated with chromic acid or other appropriate oxidizing agents.

$$R-CHO + \text{oxidation} \rightarrow R-CO-OH \rightarrow \begin{matrix}\text{oxidized} \\ \text{derivatives}\end{matrix} \rightarrow \text{cleavage}$$

$$R_2C{=}O + \text{oxidation} \rightarrow \text{oxidized derivatives} \rightarrow \text{cleavage}$$

Ex.

$$3CH_3-CHO + Cr_2O_7 = \xrightarrow{+8H^+} 3CH_3-CO-OH + 2Cr^{+++} + 4H_2O$$

$$CH_3-CO-CH_2-CH_2-CH_3 + Cr_2O_7 = \xrightarrow{+8H^+} CH_3-CO-OH$$

$$+ CH_3-CH_2-COOH$$

$$+ 2Cr^{+++} + 4H_2O$$

By reduction when treated with appropriate reducing agents (Zn/H^+, Na/ROH, $NaHg/H_2O$).

$$R-CHO + 2Na/2EtOH \rightarrow RCH_2-OH + 2Et-O-Na$$

1° alcohol

$$R_2C{=}O + 2Na/2Et-OH \rightarrow R_2CH-OH + 2Et-O-Na$$

2° alcohol

By addition when treated with a Grignard reagent, hydrogen cyanide, sodium hydrogen sulfite, or ammonia.

A) $R{-}CHO + R'{\cdot}MgX \rightarrow R{\cdot}CH(O{\cdot}MgX){\cdot}R'$

$R_2C{=}O + R'{\cdot}MgX \rightarrow R_2C(O{\cdot}MgX){\cdot}R'$

B) $R{-}CHO + H{\cdot}CN \rightarrow R{\cdot}CH(O{\cdot}H){\cdot}CN$

$R_2C = O + H{\cdot}CN \rightarrow R_2C(O{\cdot}H){\cdot}CN$

C) $R{\cdot}CHO + H{\cdot}OSO_2Na \rightarrow R{\cdot}CH(O{\cdot}H){\cdot}SO_3Na$

$R_2C = O + H{\cdot}OSO_2Na \rightarrow R_2C(O{\cdot}H){\cdot}SO_3Na$

D) $R{\cdot}CHO + H{\cdot}NH_2 \rightarrow R{\cdot}CH(O{\cdot}H){\cdot}NH_2$

$R_2C = O + H{\cdot}NH_2 \rightarrow$ complex derivatives

Substitution reactions

A) $RCH\boxed{O + 2H}OR' \rightarrow R{-}CH{-}(OR)_2' + H_2O$

$R_2C = O + 2HOR' \not\rightarrow$

B) $RCH\boxed{O + PCl_3}Cl_2 \rightarrow RCH{\cdot}Cl_2 + P{\cdot}O{\cdot}Cl_3$

$R_2C = \boxed{O + PCl_3}Cl_2 \rightarrow R_2C{\cdot}Cl_2 + P{\cdot}O{\cdot}Cl_3$

C) $R{\cdot}CO\boxed{H + X}X \rightarrow R{\cdot}CO{\cdot}X + H{\cdot}X$

$R{\cdot}CO{\cdot}HRC\boxed{H + X}X \rightarrow R{\cdot}CO{\cdot}HRC{\cdot}X + H{\cdot}X$

Condensation reaction

$R{-}CHO + R'{-}CH_2{-}CHO(Ca(OH)_2, aq.)$

$\rightarrow R{-}CHOH{-}CHR'{-}CHO$

$R{-}CHOH\ CHR'{-}CHO \xrightarrow[-H_2O]{} RHC{=}CR'{-}CHO$

Polymerization of Aldehydes

$x(H{-}CHO)aq \xrightarrow{evaporate} (H{-}CHO)_x$, paraformaldehyde

Resin formation by the presence of condensing agent

$H{-}CHO + C_6H_5OH$(Phenol) $\rightarrow$ synthetic resin "bakelite"

$R{-}CHO + NaOH$, aq., conc. $\xrightarrow{heat}$ resin formation

Reaction of aldehydes and ketones with bases:

$$\rangle \overset{\delta^+}{C} = \overset{\delta^-}{\ddot{\underset{..}{O}}}: \; + \; :\ddot{O}\langle^{H}_{H} \; \underset{\longleftarrow}{\overset{H^+}{\longrightarrow}} \; \rangle C \langle^{OH}_{OH}$$

Reaction with derivatives of ammonia:

$$\rangle C = O + (H_2N) \rightarrow \rangle C = N- \; + \; H_2O$$

carbonyl group — azomethene group

Cannizzaro reaction. In the presence of concentrated alkali, aldehydes containing no α–hydrogens undergo self-oxidation and reduction to yield an alcohol and a salt of a carboxylic acid.

$$2-\overset{\overset{H}{|}}{C}=O \xrightarrow{\text{strong base}} -COO^- + -CH_2OH$$

aldehyde with no α -hydrogens

$$\text{Ex.} \quad 2HCHO \xrightarrow{50\%\ NaOH} CH_3OH \; + \; HCOO^-Na^+$$

formaldehyde — methanol — sodium formate

Aldol condensation. In the presence of a dilute base or acid, two molecules of a ketone, or an aldehyde, containing α hydrogens combine to form an aldol (β-hydroxy ketone) or (β hydroxy aldehyde).

$$\rangle C = O + -\underset{\underset{H}{|}}{\overset{|}{C}}-\overset{|}{C} = O \xrightarrow{\text{base or acid}} -\overset{|}{\underset{|}{C}}-\overset{|}{\underset{\underset{OH}{|}}{C}}-\overset{|}{C} = O$$

an aldol

$$\text{Ex.} \quad 2CH_3\overset{H}{C}=O \xrightarrow{OH} CH_3\underset{\underset{OH}{|}}{C}HCH_2CHO$$

acetaldehyde 2 moles — acetaldol

Cyanohydrin formation by the nucleophilic addition of a cyanide anion to the carbonyl group

$$\ce{>C^{\delta+}=\ddot{O}:^{\delta-} + H^{+}\ :C#N:^{-} -> -C(-:\ddot{O}:^{-})-C#N + H^{+} -> -C(-OH)-C#N}$$

Reduction to hydrocarbons

$$\ce{>C=O ->[Zn(Hg), conc. HCl] -C(H)-H}$$

Clemmensen reduction for compounds sensitive to base

$$\ce{>C=O ->[N_2H_4\ base] -C(H)-H}$$

Wolff-Kishner reduction for compounds sensitive to acid

CHAPTER 16

AMINES

Amines are derivatives of hydrocarbons in which a hydrogen atom has been replaced by an amino group; derivatives of ammonia in which one or more hydrogen atoms have been replaced by alkyl groups also known as amines. They are classified, according to structure, as

$$\text{Primary - } R{-}\underset{\displaystyle H}{\underset{|}{N}}{-}H, \quad \text{Secondary - } R{-}\overset{\displaystyle H}{\overset{|}{N}}{-}R,$$

$$\text{Tertiary - } R{-}\overset{\displaystyle R}{\overset{|}{N}}{-}R$$

16.1 NOMENCLATURE (IUPAC SYSTEM)

The aliphatic amine is named by listing the alkyl groups attached to the nitrogen, and following these by "-amine."

$CH_3{-}NH_2$ Methyl-amine

$CH_3{-}\overset{CH_3}{\overset{|}{\underset{NH_2}{\underset{|}{C}}}}{-}CH_3$ tert.-Butyl-amine

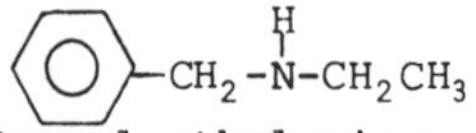

$C_6H_5{-}CH_2{-}\overset{H}{\overset{|}{N}}{-}CH_2CH_3$ Benzyl ethylamine

If an alkyl group occurs twice or three times on the nitrogen, the prefixes "di-" and "tri-" are used respectively.

Ex. $CH_3{-}NH{-}CH_3$ dimethylamine

$CH_3{-}\overset{CH_3}{\overset{|}{N}}{-}CH_3$ trimethylamine

If an amino group is part of a complicated molecule, it may be named by prefixing "amino" to the name of the parent chain.

Ex. $NH_2CH_2CH_2OH$ — 2-amino ethanol

$$CH_3\underset{}{\overset{NH_2}{\overset{|}{C}}}H\ CH_2COOH$$

3-aminobutanoic acid

An amino substituent that carries an alkyl group is named as an N-alkyl amino group.

Ex. $CH_3NH{-}CH_2COOH$ — N-methyl amino acetic acid

$$CH_3{-}NH\overset{CH_3}{\overset{|}{C}}H(CH_2)_4CH_3$$

2-(N-methylamino) heptane

16.2 PHYSICAL PROPERTIES OF AMINES

A) Amines are polar compounds.

B) Primary amines form intermolecular hydrogen bonds which are weaker than those of alcohols and carboxylic acids.

C) Amines have higher boiling points than non-polar compounds of comparable molecular size, but lower boiling points than alcohols or carboxylic acids.

D) Amines up to about six carbons are quite soluble in water and more basic than ammonia.

E) The water solubility of amines decreases with increasing size of non-polar alkyl groups attached to nitrogen.

F) Amines are soluble in less polar solvents such as alcohol, ether and benzene.

G) At room temperature, the lower members are gases, propylamine to dodecylamine are liquids, and the higher members are solids.

16.3 PREPARATION OF AMINES

Substitution reactions. Alkyl halides undergo nucleophilic substitution by ammonia yielding ammonium salts. Subsequent treatment with a base liberates the free amine.

$$X{-}CH_2R + :NH_3 \rightarrow \left[\overset{\delta^-}{X} \leftarrow \text{----} C(R)(H)(H) \leftarrow \text{----} \overset{\delta^+}{NH_3} \right] \rightarrow$$

$$R{-}CH_2{-}\overset{\oplus}{N}H_3X^{\ominus}$$

$$R{-}CH_2{-}\overset{\oplus}{N}H_3X^{\ominus} + OH^- \rightarrow R{-}CH_2NH_2 + H_2O + X^{\ominus}$$

The reduction of appropriate compounds by catalytic hydrogenation or use of certain reducing agents:

A) a) $R{-}CN + 4Na/4CH_3{-}CH_2{-}OH \rightarrow R{-}CH_2{-}NH_2 + 4CH_3{-}CH_2{-}ONa$

b) $R{-}C(=NOH){-}H + 4Na/4CH_3{-}CH_2{-}OH \rightarrow R{-}CH_2{-}NH_2 + 4CH_3{-}CH_2{-}O{-}Na + H_2O$

c) $R{-}C(=NNH_2){-}H + 4Na/4CH_3{-}CH_2{-}OH \rightarrow R{-}CH_2{-}NH_2 + 4CH_3{-}CH_2{-}O{-}Na + NH_3$

d) $4R{-}CH_2{-}NO_2 + 9Fe/(FeCl_2, H^+) + 4H_2O \rightarrow 4R{-}CH_2{-}NH_2 + 3Fe_3O_4$

B) $R{-}NC + 4Na/4CH_3{-}CH_2{-}OH \rightarrow R{-}NH{-}CH_3 + 4CH_3{-}CH_2{-}O{-}Na$

Hofmann reaction. Reduction of amides by bromine and alkali to give primary amines.

$R{-}CO{-}NH_2 + Br_2/NaOH, aq \rightarrow R{-}CO{-}NHBr + NaBr + H_2O$

$R{-}CO{-}NHBr + NaOH, aq \rightarrow R{-}N{=}C{=}O + NaBr + H_2O$

alkyl isocyanate

$R{-}N{=}C{=}O + 2NaOH, aq \rightarrow R{-}NH_2 + Na_2CO_3$

primary
amine

Gabriel synthesis. Reaction of alkyl halides and alkali with potassium phthalimide to give primary amines.

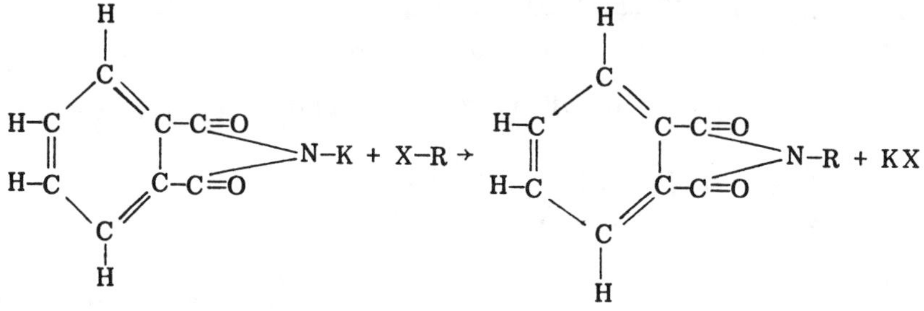

```
       H                                    H
       |                                    |
       C                                    C
     /   \\                               /   \\
H-C        C-C=O                  H-C        C-C=O
  ||       |      >N-K + X-R →      ||       |      >N-R + KX
H-C        C-C=O                  H-C        C-C=O
     \   //                               \   //
       C                                    C
       |                                    |
       H                                    H
```

alkyl phthalimide

```
       H                                      H
       |                                      |
       C                                      C
     /   \\                                 /   \\
H-C        C-C=O                    H-C        C-CO-ONa
  ||       |      >N-R + 2NaOH →      ||       |            + R-NH2
H-C        C-C=O                    H-C        C-CO-ONa
     \   //                                 \   //          primary
       C                                      C             amine
       |                                      |
       H                                      H
```

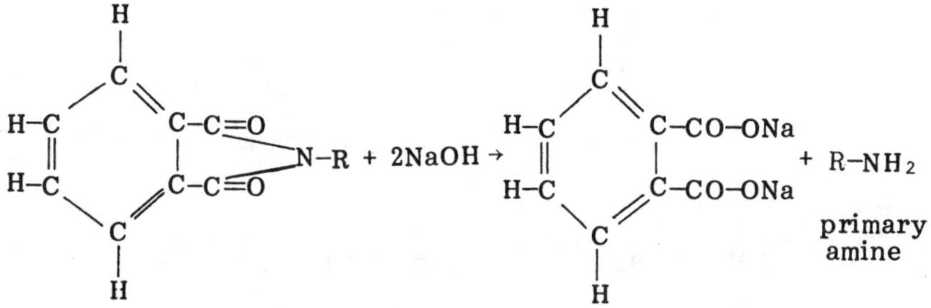

Reaction of methanol with ammonia in the presence of a catalyst, to give a mixture of methyl amines.

$CH_3OH + NH_3/\text{Catalyst} \rightarrow CH_3{-}NH_2 + H_2O$

$CH_3OH + CH_3{-}NH_2/\text{Catalyst} \rightarrow (CH_3)_2NH + H_2O$

$CH_3OH + (CH_3)_2{-}NH/\text{Catalyst} \rightarrow (CH_3)_3N + H_2O$

Ethanolamines are prepared by the action of ammonia on ethylene oxide.

```
       O
      / \
A) H2C---CH2 + NH3 →  HO-CH2-CH2-NH2  (ethanolamine)
```

B) $2H_2C\overset{O}{—}CH_2 + NH_3 \rightarrow (HO{-}CH_2{-}CH_2)_2N{-}H$ (diethanolamine)

C) $3H_2C\overset{O}{—}CH_2 + NH_3 \rightarrow (HO{-}CH_2{-}CH_2)_3N$ (triethanolamine)

The reduction of nitriles by hydrogen and a catalyst to produce primary amines.

$$RC \equiv N \xrightarrow{2H_2,\ Catalyst} \underset{\text{primary amine}}{R{-}CH_2{-}NH_2}$$

The reduction of nitro compounds to give primary amines:

$$\begin{matrix} ArNO_2 \\ or \\ RNO_2 \end{matrix} \xrightarrow[H_2,\ catalyst]{metal,\ H^+;\ or} \begin{matrix} ArNH_2 \\ or \\ RNH_2 \end{matrix}$$

A) Catalytic hydrogenation

Ex. $p\text{-}NO_2{-}C_6H_4{-}COOCH_3 \xrightarrow{H_2/Pd} p\text{-}NH_2{-}C_6H_4{-}COOCH_3$

Methyl-p-nitrobenzoate → Methyl-p-aminobenzoate

B) Metal-acid reduction

Ex. $CH_3CH_2CH_2{-}NO_2 \xrightarrow{Fe/HCl} CH_3CH_2CH_2NH_2$

1-nitropropane → n-propylamine

C) Lithium aluminum hydride

Ex. $CH_3CH_2CH(NO_2)CH_3 \rightarrow CH_3CH_2CH(NH_2)CH_3$

2-nitrobutane → 2-butylamine

16.4 REACTIONS OF AMINES

FORMATION OF ADDITION PRODUCTS

A) 1) $R{-}NH_2 + HX \rightarrow R{-}NH_2{-}HX$

2) $R-NH_2 + R-X \rightarrow R_2-N-H-HX$

B) 1) $R_2NH + H-X \rightarrow R_2NH-HX$

2) $R_2NH + R-X \rightarrow R_3N-HX$

C) 1) $R_3N + H-X \rightarrow R_3N-HX$

2) $R_3N + R-X \rightarrow R_4N-X$

Simultaneous replacement of both hydrogen atoms of the NH_2 group

$$RNH_2 + CHCl_3 \rightarrow R-NC + 3HCl$$

CONVERSION INTO AMIDES

A) RNH_2 —

$$\xrightarrow[-HCl]{R'COCl} R'CONH \cdot R \quad \text{(N-substituted amide)}$$

$$\xrightarrow[-HCl]{ArSO_2Cl} ArSO_2NHR \quad \text{(N-substituted sulfonamide)}$$

B) R_2NH —

$$\xrightarrow[-HCl]{R'COCl} R'CONR_2 \quad \text{(N,N-disubstituted amide)}$$

$$\xrightarrow[ArSO_2Cl \; -HCl]{} ArSO_2NR_2 \quad \text{(N,N-disubstituted sulfonamide)}$$

C) Tertiary amines, R_3N, do not react in this manner.

FORMATION OF AMMONIUM BASES

A) $R-NH_2 + H_2O \rightleftharpoons (R-NH_3 - OH) \rightleftharpoons R-NH_3^{\oplus} + OH^{\ominus}$

B) $R_2-NH + H_2O \rightleftharpoons (R_2NH_2-OH) \rightleftharpoons R_2-NH_2^{\oplus} + OH^{\ominus}$

C) $R_3N + H_2O \rightleftharpoons (R_3NH-OH) \rightleftharpoons R_3-NH^{\oplus} + OH^{\ominus}$

Halogenation of amines by hypochlorous acid or tert-butyl-hypobromite in alkaline solution:

A) $R{-}NH_2 + 2Cl_2 \xrightarrow{NaOH/Cl_2} R{-}N\begin{matrix} Cl \\ Cl \end{matrix} + 2HCl$

N,N-dichloroalkylamine

B) $R_2NH + Cl_2 \xrightarrow{NaOH/Cl_2} R_2N{-}Cl + HCl$

N–Chlorodialkylamine

Basicity of amines: since ammonia and amines contain a nitrogen with an unshared electron pair, they act as bases, accepting protons and forming ammonium ions and alkyl ammonium ions.

A) $\ddot{N}H_3 + H^{\oplus} \rightarrow NH_4^{\oplus}$, Ammonium ion

B) $R{-}\ddot{N}H_2 + H^{\oplus} \rightarrow R{-}NH_3^{\oplus}$, Alkylammonium ions

Ammonia and amines are stronger bases than water, forming ammonium salts in aqueous mineral acids.

REACTION OF AMINES WITH NITROUS ACID

A) Primary aliphatic and aromatic amines are converted into diazonium salts.

a) $ArNH_2 \xrightarrow{HONO} ArN \equiv N^{\oplus}$

aryldiazonium salt

b) $RNH_2 \xrightarrow{HONO} [R{-}N \equiv N^{\oplus}] \xrightarrow{H_2O} N_2$ + mixture of alcohols and alkenes

alkyldiazonium salt unstable

B) Secondary amines react to yield N-nitrosoamines

$ArNHR$ or $R_2NH \xrightarrow{HONO} Ar{-}N(R){-}N = O$ or $R_2N{-}N = O$

C) Tertiary aliphatic amines are oxidized to yield N-nitrosodialkylamines and a mixture of ketones and aldehydes.

$R_3N \xrightarrow{HONO} R_2N{-}N{=}O$ + mixture of aldehydes and ketones

D) Tertiary aromatic amines undergo electrophilic nitrosation at the benzene ring.

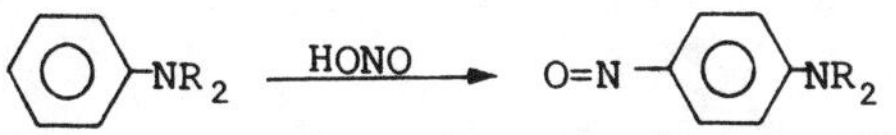

$$C_6H_5-NR_2 \xrightarrow{HONO} O{=}N-C_6H_4-NR_2$$

p-nitroso compound

Electrophilic substitution of aromatic amines. Amino groups ($-NH_2$, $-NHR$, $-NR_2$), but not ammonium groups in aryl ammonium ions, activate the benzene ring to which they are attached to for electrophilic substitution. They release electrons due to their unshared electron pairs.

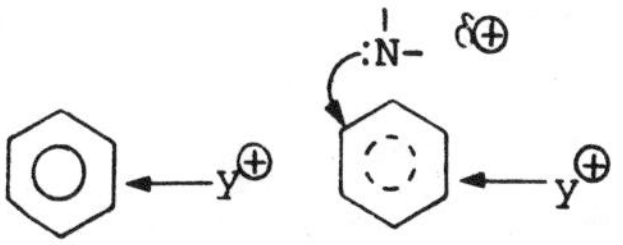

:N− δ⊕

$C_6H_6 \leftarrow Y^{\oplus}$ $\leftarrow Y^{\oplus}$

16.5 AROMATIC DIAZONIUM SALTS

Aliphatic diazonium salts are very unstable and decompose spontaneously after generation. Aromatic diazonium salts, on the other hand, are more stable and are very reactive. They are important intermediates and are seldom isolated or purified.

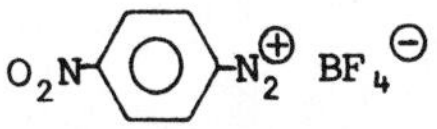

$O_2N-C_6H_4-N_2^{\oplus}\ BF_4^{\ominus}$ p-Nitrobenzene diazonium fluoroborate

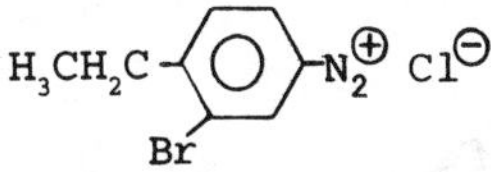

$H_3CH_2C-C_6H_3(Br)-N_2^{\oplus}\ Cl^{\ominus}$ 3-Bromo-4-ethyl benzene diazonium chloride

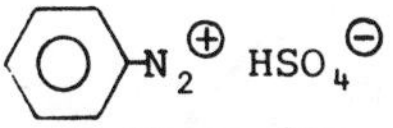

$C_6H_5-N_2^{\oplus}\ HSO_4^{\ominus}$ Benzene diazonium hydrogen sulfate

NOMENCLATURE

Compounds containing the group $-N{\equiv}N^{\oplus}$ are called diazonium salts. When naming aromatic diazonium salts, the word "diazonium" is added to the name of the aryl group, followed by the name of the anion.

REACTIONS OF AROMATIC DIAZONIUM SALTS

A) Nucleophilic displacement of nitrogen.

$$ArN_2^{\oplus} + :Z \rightarrow ArZ + N_2$$

a) Replacement by hydroxyl.

$$ArN_2^{\oplus} + H:OH \xrightarrow{H^+} ArOH + N_2$$
phenol

$$o\text{-}CH_3C_6H_4N_2^{\oplus}HSO_4^{\ominus} + H_2O \xrightarrow[heat]{H^+} o\text{-}CH_3C_6H_4OH + N_2 + H_2SO_4$$
o-Cresol

b) Replacement by hydrogen.

$$ArN_2^{\oplus} + HO:H \xrightarrow{H_3PO_2} ArH + H_3PO_3 + N_2$$

$$2,4\text{-}Cl_2C_6H_3N_2^{\oplus}HSO_4^{\ominus} + H_2O \xrightarrow{H_3PO_2} m\text{-}C_6H_4Cl_2 + H_3PO_3 + H_2SO_4 + N_2$$
m-Dichlorobenzene

c) Replacement by iodine.

$$ArN_2^{\oplus} + :I^{\ominus} \longrightarrow ArI + N_2$$

$$C_6H_5N_2^{\oplus}HSO_4^{\ominus} + K:I \longrightarrow C_6H_5I + N_2 + KHSO_4$$
Iodobenzene

d) Replacement by fluorine.

$$ArN_2^{\oplus}BF_4^{\ominus} \xrightarrow{heat} ArF + N_2 + BF_3$$

$$C_6H_5N_2^{\oplus}BF_4^{\ominus} \xrightarrow{heat} C_6H_5F + N_2 + BF_3$$
Fluorobenzene

e) Replacement by chlorine, bromine, or the nitrile group. Sandmeyer reaction.

1) $ArN_2^{\oplus} + Cu:Cl \rightarrow ArCl + N_2$

2) $ArN_2^{\oplus} + Cu{:}Br \rightarrow ArBr + N_2$

3) $ArN_2^{\oplus} + Cu{:}CN \rightarrow ArCN + N_2$

$$o\text{-}CH_3C_6H_4N_2^{\oplus}Cl^{\ominus} + CuCl \longrightarrow o\text{-}CH_3C_6H_4Cl + N_2 + CuCl$$

o-Chlorotoluene

$$o\text{-}CH_3C_6H_4N_2^{\oplus}HSO_4^{\ominus} + CuBr \longrightarrow o\text{-}CH_3C_6H_4Br + N_2 + CuHSO_4$$

o-Bromotoluene

$$o\text{-}CH_3C_6H_4N_2^{\oplus}Cl^{\ominus} + CuCN \longrightarrow o\text{-}CH_3C_6H_4CN + N_2 + CuCl$$

o-Tolunitrile

f) Replacement by the thiol group (–SH).

Ex.

$$C_6H_5N_2^{\oplus}HSO_4^{\ominus} + EtO\text{-}\overset{S}{\overset{\|}{C}}\text{-}SK \longrightarrow C_6H_5S\text{-}\overset{S}{\overset{\|}{C}}\text{-}CEt \xrightarrow{KOH,\,hyd}$$

potassium ethyl xanthate

$$EtOH + COS + C_6H_5S^{\ominus}K^{\oplus} \xrightarrow{H_3O^+} C_6H_5SH$$

g) Replacement by the nitro group.

Ex.

$$C_6H_5N_2^{\oplus}BF_4^{\ominus} + NaNO_2 \xrightarrow{Cu} C_6H_5NO_2 + NaBF_4 + N_2$$

h) The Gatterman reaction. Replacement by aryl group.

Ex.

$$2C_6H_5N_2^{\oplus}HSO_4^{\ominus} \xrightarrow{Cu} C_6H_5\text{—}C_6H_5$$

Biphenyl

B) Reduction to hydrazines

Ex.

$$C_6H_5N_2^{\oplus}Cl^{\ominus} \xrightarrow[H_2O]{Na_2SO_3} C_6H_5NHNH_2$$

C) Coupling

$$ArN_2^{\oplus} X^{\ominus} + C_6H_5\text{-}G \longrightarrow Ar\text{-}N{=}N\text{-}C_6H_4\text{-}G$$

An azo compound

G must be a strong electron-releasing group:

OH, NR_2, NHR, NH_2

$$C_6H_5N_2^{\oplus}Cl^{\ominus} + C_6H_5\text{-}OH \xrightarrow{\text{weakly alkaline}} C_6H_5\text{-}N{=}N\text{-}C_6H_4\text{-}OH + HCl$$

$$Me_2N\text{-}C_6H_5 + C_6H_5\text{-}\overset{\oplus}{N_2}\overset{\ominus}{Cl} \longrightarrow Me_2N\text{-}C_6H_4\text{-}N{=}N\text{-}C_6H_5 + HCl$$

CHAPTER 17

PHENOLS AND QUINONES

17.1 NOMENCLATURE OF PHENOLS

Phenols have the general formula ArOH. The -OH group in phenols is attached directly to the aromatic ring.

Phenols are named as derivatives of phenol, which is the simplest member of the family. Methyl phenols are given the name cresols. Phenols are also called "hydroxy-" compounds.

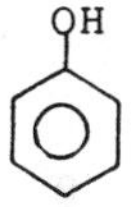

Phenol 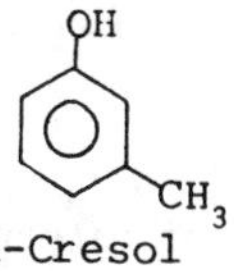   m-Cresol 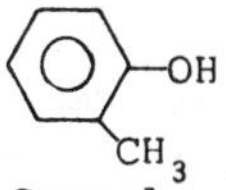   o-Cresol

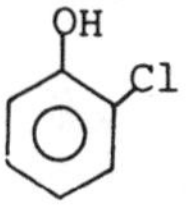

o-Chlorophenol 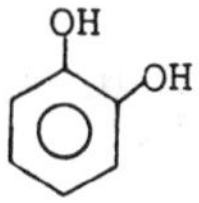   Catechol 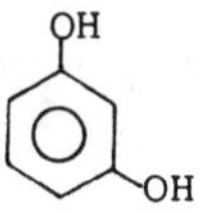   Resorcinol    Hydroquinone

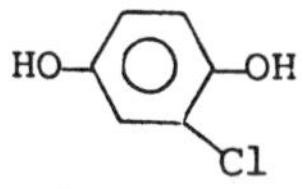

2-Chlorohydroquinone 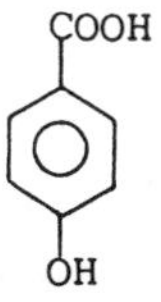   p-Hydroxybenzoic acid   Picric acid

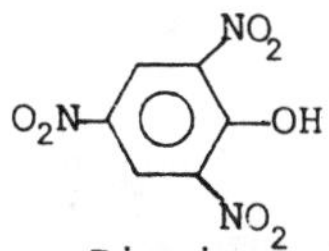

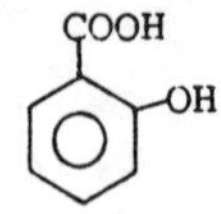

Salicylic acid

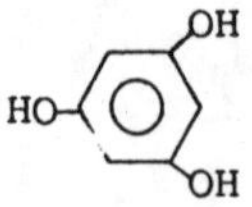

Phloroglucinol

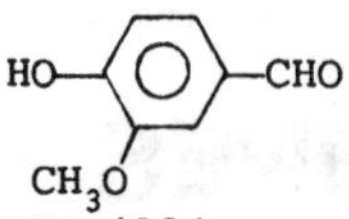

Vanillin

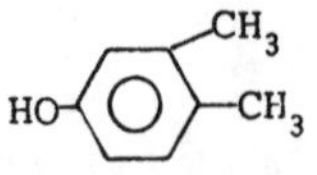

3,4-Xylenol

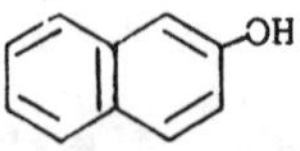

β-Naphthol (or 2-Naphthol)

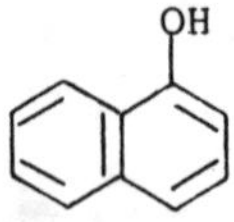

α-Naphthol

17.2 PHYSICAL PROPERTIES OF PHENOLS

Simple phenols are low-melting solids or liquids. They have high boiling points because of hydrogen bonding. Phenol (C_6H_5OH) is soluble in water to some extent because of the hydrogen bonding present in both. Most phenols are insoluble in water. Phenols are colorless. They are easily oxidized, except when purified. Many phenols are colored by oxidation products.

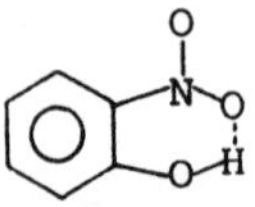

Intramolecular hydrogen bonding: Chelation

The intramolecular hydrogen bonding within a single molecule, takes the place of intermolecular hydrogen bonding with other phenol molecules and with water molecules.

The holding of a hydrogen or metal atom between two atoms of a single molecule is called chelation.

Acidity of Phenols

$$ArOH \underset{H^+}{\overset{OH^-}{\rightleftharpoons}} ArO^-$$

a phenol (acid)	a phenoxide ion (salt)
insoluble in water	soluble in water

The solubility properties of phenols and their salts are opposite, the salts are insoluble in organic solvents and soluble in water.

Phenols are stronger acids than alcohols because the negative charge created by deprotonation can be distributed in the phenol by resonance. Deprotonation lowers the energy level of the compound and also stabilizes it.

Because phenols dissolve in aqueous hydroxide, but not in aqueous bicarbonate, they are more acidic than water, but less acidic than carboxylic acid.

17.3 PREPARATION OF PHENOLS

A) Nucleophilic displacement of halides

$$C_6H_5Cl \xrightarrow[4500\ lb/in^2]{NaOH,\ 360^\circ} C_6H_5O^{\ominus}Na^{\oplus} \xrightarrow{HCl} C_6H_5OH + NaCl$$

Chlorobenzene   Sodium phenoxide   Phenol

$$O_2N{-}C_6H_4{-}Cl + OH^- \rightarrow O_2N{-}C_6H_4{-}OH + Cl^-$$

p–Nitrochlorophenol   p-Nitrophenol

B) Oxidation of cumene

$$C_6H_5C(CH_3)_2H \xrightarrow{O_2} C_6H_5C(CH_3)_2OOH \xrightarrow{H_2O,\ H^+} C_6H_5OH + H_3C{-}C(=O){-}CH_3$$

Cumene   Cumene hydroperoxide   Phenol   Acetone

C) Hydrolysis of diazonium salts

$$Ar\overset{+}{N}_2 + H_2O \rightarrow ArOH + H^+ + N_2$$

$N_2^{\oplus}HSO_4^{\ominus}$, Cl → ($H_2O, H^+$, heat) → OH, Cl + N_2

m-Chlorobenzene-diazonium hydrogen sulfate   m-Chlorophenol

D) Oxidation of organothallium compounds

o-Xylene → ($Tl(OOCCF_3)_3$) → $Tl(OOCCF_3)_2$ → ($Pb(OAc)_4$) → $OOCCF_3$ → (dil.NaOH, H_2O) → OH + CF_3COONa

3,4-Xylenol

E) Oxidation of arylthallium compounds

$$ArTl(OOCCF_3)_2 \xrightarrow{Pb(OAc)_4} ArOOCCF_3 \xrightarrow[heat]{H_2O, OH^-} ArO^- \xrightarrow{H^+} ArOH$$

arylthallium trifluoroacetate   trifluoro-acetate

$$ArH \xrightarrow{Tl(OOCCF_3)_3} ArTl(OOCCF_3)_2$$

Chlorobenzene → ($Tl(OOCCF_3)_3$) → ($Pb(OAc)$, Ph_3P) → (H_2O, OH^-, heat) → (H^+) → p-Chlorophenol

Sodiumbenzenesulfonate ($SO_3^{\ominus}Na^{\oplus}$) → (NaOH, fuse) → O^-Na^+ → (H_3O^+) → Phenol

17.4 REACTIONS OF PHENOLS

A) Salt formation

$$ArOH + H_2O \rightleftarrows ArO^- + H_3O^+$$

$$C_6H_5{-}OH + NaOH \longrightarrow C_6H_5{-}O^{\ominus}Na^{\oplus} + H_2O$$

Phenol — Sodiumphenoxide

B) Ether formation-Williamson synthesis

$$ArOH \xrightarrow{OH^-} ArO^- \begin{cases} \xrightarrow{RX} Ar{-}O{-}R + X^- \\ \xrightarrow{(CH_3)_2SO_4} Ar{-}O{-}CH_3 + CH_3SO_4^{\ominus} \end{cases}$$

$$H_3C{-}C_6H_4{-}OH + BrH_2C{-}C_6H_4{-}NO_2 \xrightarrow[\text{heat}]{\text{Aqueous NaOH}} H_3C{-}C_6H_4{-}O{-}CH_2{-}C_6H_4{-}NO_2$$

p-Cresol — p-Nitrobenzyl bromide — p-Tolyl p-nitrobenzyl ether

$$o\text{-}NO_2C_6H_4OH + (CH_3)_2SO_4 \xrightarrow[\text{heat}]{\text{Aq.NaOH}} o\text{-}NO_2C_6H_4OCH_3 + CH_3SO_4Na$$

o-Nitrophenol — Methylsulfate — o-Nitroanisole (o-Nitrophenyl methyl ether)

$$C_6H_5{-}OH + ClCH_2COOH \xrightarrow[\text{heat}]{\text{Aq.NaOH}} C_6H_5{-}OCH_2COONa \xrightarrow{HCl} C_6H_5{-}OCH_2COOH$$

Phenol — Chloroacetic acid — Phenoxyacetic acid

C) Ester formation

$$HOAc + ArOAc \xleftarrow{Ac_2O} ArOH \xrightarrow[(ArCOCl)]{RCOCl} ArOOCR(ArOOCAr) + HCl$$

$$TlCl + ArOTs \xleftarrow{TsCl} ArOTl \xrightarrow[ArCOCl]{RCOCl} ArOOCR(ArOOCAr) + TlCl$$

$$ArOH \longrightarrow \begin{cases} \xrightarrow{RCOCl\ (ArCOCl)} RCOOAr(ArCOOAr) \\ \xrightarrow{Ar'SO_2Cl} Ar'SO_2OAr \end{cases}$$

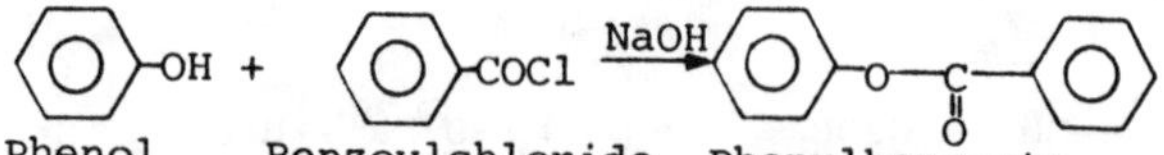

Phenol + Benzoylchloride $\xrightarrow{NaOH}$ Phenylbenzoate

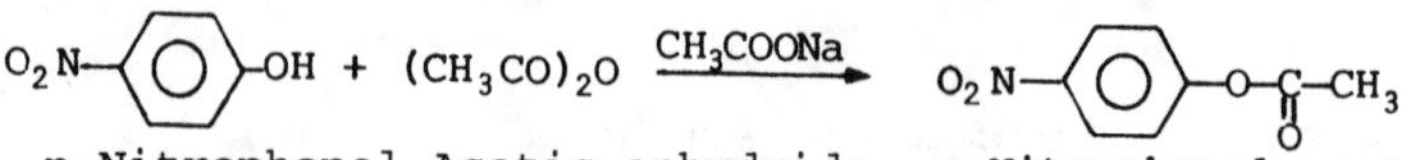

p-Nitrophenol + Acetic anhydride $(CH_3CO)_2O$ $\xrightarrow{CH_3COONa}$ p-Nitrophenyl acetate

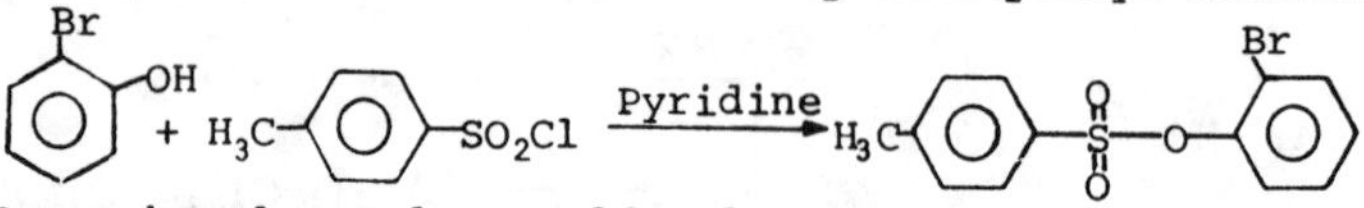

o-Bromophenol + p-Toluenesulfonyl chloride $\xrightarrow{Pyridine}$ o-Bromophenyl-p-Toluene sulfonate

D) Ring Substitution

-OH Activate powerfully, and direct ortho, para in
$-O^{\ominus}$ Electrophilic aromatic substitution
-OR: Less powerful activator than -OH.

a) Nitration

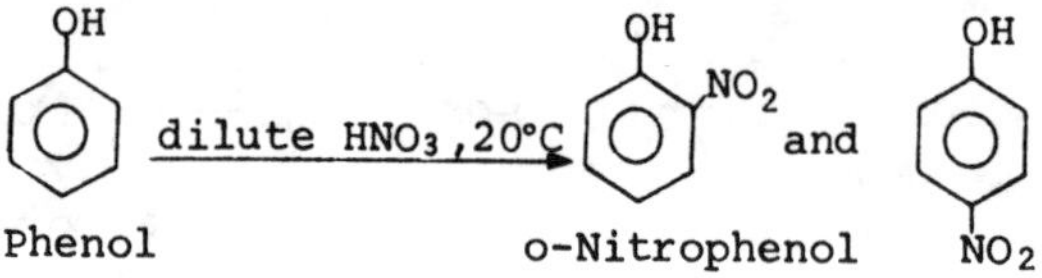

Phenol $\xrightarrow{dilute\ HNO_3,\ 20°C}$ o-Nitrophenol and p-Nitrophenol

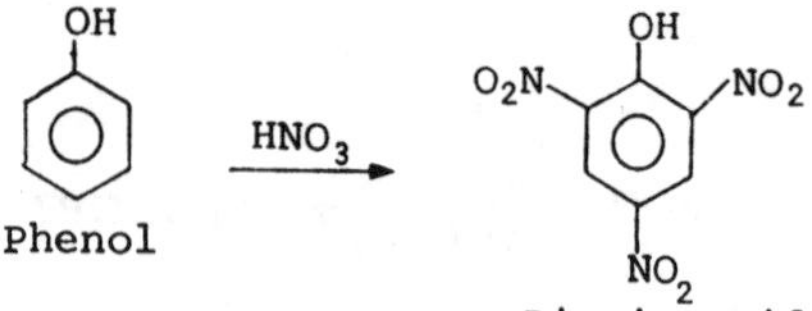

Phenol $\xrightarrow{HNO_3}$ Picric acid

b) Sulfonation

$$\text{Phenol} \xrightarrow{H_2SO_4} \begin{cases} \xrightarrow{15\text{-}20^\circ C} \text{o-Phenolsulfonic acid} \xrightarrow{H_2SO_4,\ 100^\circ C} \text{p-Phenolsulfonic acid} \\ \xrightarrow{100^\circ C} \text{p-Phenolsulfonic acid} \end{cases}$$

o-Phenolsulfonic acid

p-Phenolsulfonic acid

c) **Halogenation**

$$\text{Phenol} \xrightarrow{Br_2,\ H_2O} \text{2,4,6-Tribromophenol}$$

$$\text{Phenol} \xrightarrow{Br_2,\ Cs_2,\ 0^\circ C} \text{p-Bromophenol}$$

Phenol

2,4,6-Tribromophenol

Phenol

p-Bromophenol

d) **Nitrosation**

$$\text{o-Cresol} + NaNO_2 + H_2SO_4 \longrightarrow \text{4-Nitroso-2-methylphenol}$$

o-Cresol

4-Nitroso-2-methylphenol

e) **Friedel-Crafts alkylation**

$$\text{Phenol} + CH_3-C(CH_3)(Cl)-CH_3 \xrightarrow{HF} \text{p-tert-Butylphenol}$$

Phenol

tert-Butyl chloride

p-tert-Butylphenol

C_6H_5OH $\xrightarrow{HONO}$ ON–C_6H_4–OH $\xrightarrow{HNO_3, Ox.}$ O_2N–C_6H_4–OH

Phenol — p-Nitrosophenol — p-Nitrophenol

f) **Friedel-Crafts acylation. Fries rearrangement.**

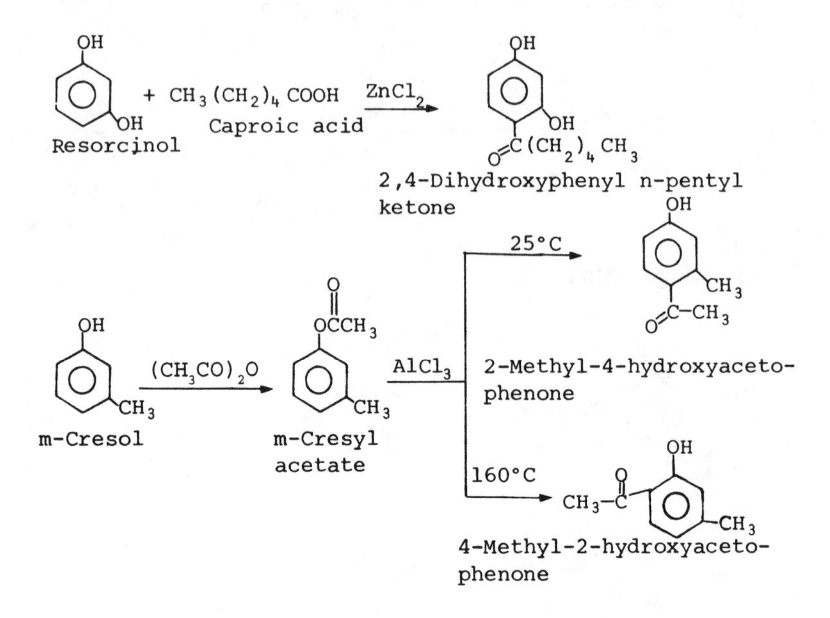

g) **Coupling with diazonium salts**

HO–C_6H_5 + $C_6H_5N_2Cl$ ⟶ HO–C_6H_4–N=N–C_6H_5 + HCl

Phenol — Benzenediazonium chloride — p-Hydroxyazobenzene (an azo compound)

h) **Carbonation. Kolbe reaction.**

C_6H_5ONa + CO_2 $\xrightarrow{125°C, 4\text{-}7\ atm.}$ $C_6H_4(OH)COONa$

Sodiumphenoxide — Sodiumsalicylate

i) **Reaction with formaldehyde.**

Phenol $\xrightarrow[H^+ \text{ or } OH^-]{HCHO}$ o-Hydroxymethylphenol $\xrightarrow{C_6H_5OH}$ $\xrightarrow{CHCHO, C_6H_5OH}$ A phenol-formaldehyde resin

j) Introduction of the -COR group. Ketone formation.

1) Fries rearrangement

Phenyl ester $\xrightarrow[heat]{AlCl_3}$ Acyl derivatives (ketones)

2) Houben-Hoesch synthesis

Resorcinol + RCN + HCl (A nitrile) $\xrightarrow{ZnCl_2}\xrightarrow{H_2O}$ Acyl derivatives (ketones)

k) Introduction of the -CHO group (formylation). Aldehyde formation.

1) Reimer-Tiemann reaction

C_6H_5OH (Phenol) $\xrightarrow{CHCl_3, NaOH(aq.)}$ Salicylaldehyde

2) Gatterman reaction. Special case of the Houben-Hoesh reaction.

$$HO\text{-}C_6H_3(CH_3)\text{ (o-Cresol)} + HCN \xrightarrow{AlCl_3, HCl} \xrightarrow{H_2O} HO\text{-}C_6H_3(CH_3)\text{-}CHO$$

o-Cresol

3-Methyl-4-hydroxy benzaldehyde

l) Introduction of the -COOH group (carboxylation) Acid formation

1) Kolbe synthesis

$$C_6H_5OH + CO_2 \xrightarrow[\text{pressure}]{140°C, NaOH} \xrightarrow{H^+} o\text{-}HOC_6H_4COOH$$

Phenol

Salicylic acid

2) Use of CCl_4 in the Reimer-Tiemann reaction.

$$C_6H_5OH \xrightarrow{CCl_4, NaOH} \xrightarrow{H^+} o\text{-}HOC_6H_4COOH$$

Phenol

Salicylic acid

m) Claisen rearrangement. Allyl ethers of phenol yield o–allyl phenol upon heating

$$C_6H_5\text{-}OCH_2CH{=}CH_2 \xrightarrow{heat} o\text{-}(CH_2CH{=}CH_2)C_6H_4\text{-}OH$$

Phenylallyl ether

o-Allylphenol

n) Oxidation

$$H_3C\text{-}C_6H_4\text{-}OH \xrightarrow{Ac_2O} H_3C\text{-}C_6H_4\text{-}OAc \xrightarrow{KMnO_4, H^+} HOOC\text{-}C_6H_4\text{-}OH$$

p-Cresol

p-Hydroxybenzoic acid

17.5 QUINONES

NOMENCLATURE

Quinones are cyclic, conjugated diketones named after the parent hydrocarbon.

o-Benzoquinone p-Benzoquinone 1,4-Naphthoquinone Toluquinone (2-Methyl-1,4-benzoquinone)

PROPERTIES

Quinones are colored crystalline compounds. Since quinones are highly conjugated, they are closely balanced, energetically, against the corresponding hydroquinones. Quinones exhibit the properties of unsaturated cyclic ketones.

Preparation of Quinones

1- Phenol $\xrightarrow{+O_2}$ p-Benzoquinone $+ H_2O$

Hydroquinone $\xrightarrow{+\frac{1}{2}O_2}$ p-Benzoquinone $+ H_2O$

2- Hydroquinone $+ 2Ag^{\oplus} + 2OH^{\ominus} \longrightarrow$ p-Benzoquinone $+ 2Ag + 2H_2O$

3- Hydroquinone $\xrightarrow{FeCl_3}$ p-Benzoquinone

4 p-aminophenol $\xrightarrow[H_2SO_4]{K_2Cr_2O_7}$ $HN=C_6H_4=O$ $\xrightarrow{H_2O}$ p-Benzoquinone

5- Hydroquinone $\xrightarrow[H_2SO_4,\ 30°C]{Na_2Cr_2O_7}$ p-Benzoquinone + H_2O

Hydroquinone p-Benzoquinone

6- Catechol + Ag_2O $\xrightarrow[Na_2SO_4]{\text{anhydrous ether}}$ + H_2O + 2Ag

Catechol

7- $\xrightarrow[H_2SO_4]{Na_2Cr_2O_7}$

3-Chloro-4-aminophenol 2-Chloro-1,4-benzoquinone

8- $\xrightarrow[\text{oxidation}]{-H_2O}$ $\xrightarrow[-2NH_3]{+2H_2O}$

p-Diaminobenzene p-Quinonedimine p-Benzoquinone

9- $\xrightarrow{H_2O_2, PbO_2/\text{benzene}}$ + H_2O

2,6-Dihydroxynaphthalene 2,6-Naphthoquinone

10- 1-Naphthol $\xrightarrow[\text{eq.NaOH}]{C_6H_5N_2^{\oplus}Cl^{\ominus}}$ $N = N - C_6H_5$ $\xrightarrow[Na_2S_2O_4]{NaOH}$ HCl → $NH_3^{\oplus}Cl^{\ominus}$ $\xrightarrow[K_2Cr_2O_7]{H_2SO_4}$ 1,4-Naphthoquinone

1-Naphthol

1,4-Naphthoquinone

11- Naphthalene + $2CrO_3$ $\xrightarrow{H_2SO_4}$ Cr_2O_3 + H_2O

Naphthalene 1,4-Naphthoquinone

12- Aniline $\xrightarrow{Cr_2O_7^{--}}$ p-Benzoquinone $\underset{\text{oxidation (eg. } Fe^{++})}{\overset{\text{reduction (eg. } SO_3^{--})}{\rightleftharpoons}}$ Hydroquinone

Friedel-Crafts Acylation

13- Phthalic anhydride + benzene $\xrightarrow{AlCl_3}$ 9,10-Anthraquinone + H_2O

The preparation of quinones is achieved by oxidation of aromatic hydroxy and amino compounds and by Friedel-Crafts acylation.

REACTIONS OF QUINONES

A) Friedel-Crafts reactions

Toluene + Phthalicanhydride $\xrightarrow{AlCl_3}$ $\xrightarrow[\text{or HF}]{H_2SO_4}$ methylanthraquinone $\xrightarrow[\text{heat}]{H_2NNH_2,\ CH_3ONa}$ $\xrightarrow[\text{heat}]{\text{Pt, or Se}}$ 3-Methylanthracene

B) Diels-Alder reactions

p-Benzoquinone + 1,3 Butadiene $\longrightarrow$ $\xrightarrow{+\ \text{butadiene}}$ $\xrightarrow{Pt}$ 9,10 Dihydroxy-anthracene

C) Oxidation-reduction reactions

p-Benzoquinone + $2H^{\oplus}$ ⇌ (reduction $+2e^{\ominus}$ / $-2e^{\ominus}$ oxidation) Hydroquinone (OH, OH)

p-Benzoquinone Hydroquinone

p-Benzoquinone $\xrightarrow[ZnCl_2]{CH_3OH}$ 2,5Dimethoxy-1,4-benzoquinone (OCH_3, CH_3O) + 2 Hydroquinone (OH)

p-Benzoquinone 2,5Dimethoxy-1,4-benzoquinone Hydroquinone

D) Addition of halogen - indirect substitution

2-Methyl-1,4-naphthoquinone (CH_3) $\xrightarrow{Br_2}$ (CH_3, Br, Br, H) $\xrightarrow{AcO^-}$ 3-bromo-2-Methyl-1,4-naphthoquinone (CH_3, Br)

2-Methyl-1,4-naphthoquinone 3-bromo-2-Methyl-1,4-naphthoquinone

E) Acid-catalyzed Michael addition

p-Benzoquinone $\xrightarrow{HCl}$ 2-Chlorohydroquinone (OH, Cl, OH)

p-Benzoquinone 2-Chlorohydroquinone

F) Addition of acetic acid

p-Benzoquinone $\xrightarrow[H_2SO_4,\ 50°C]{(CH_3CO)_2O}$ 1,2,4-Triacetotoxybenzene ($OCOCH_3$, $OCOCH_3$, $OCOCH_3$)

p-Benzoquinone 1,2,4-Triacetotoxybenzene

G) Reactions of keto group

Quinones undergo most reactions which are characteristic of ketones.

$$\text{p-Benzoquinone} + 2H_2N-OH \text{ (Hydroxylamine)} \longrightarrow \text{p-Benzoquinone dioxime} + 2H_2O$$

H) Addition reactions

Quinones undergo addition at the double bond by reagents such as bromine and dienes.

$$\text{p-Benzoquinone} \xrightarrow{+Br_2} \text{2,3,5,6-tetrabromo-1,4-cyclohexadione}$$

I) 1,4-Additions

$$O{=}C_6H_4{=}O + HCl \xrightarrow{CHCl_3} \left[HO{-}C_6H_3(H)(Cl){=}O \right] \longrightarrow HO{-}C_6H_3Cl{-}OH$$

p-Benzoquinone → 2-Chlorohydroquinone

CHAPTER 18

ORGANOMETALLIC COMPOUNDS

Organometallic compounds are defined as compounds which possess direct carbon-metal bonds. This excludes salts of organic acids, metal amines and Lewis acid complexes on heteroatoms of organic molecules.

18.1 NOMENCLATURE

Organometallic compounds are named by prefixing the name of the metal with the appropriate organic radical name. A rough classification of different types of organometallic compounds can be given by the periodic table of elements given in Table 18.1.

TABLE 18.1 Characteristics of Carbon-metal Bonds

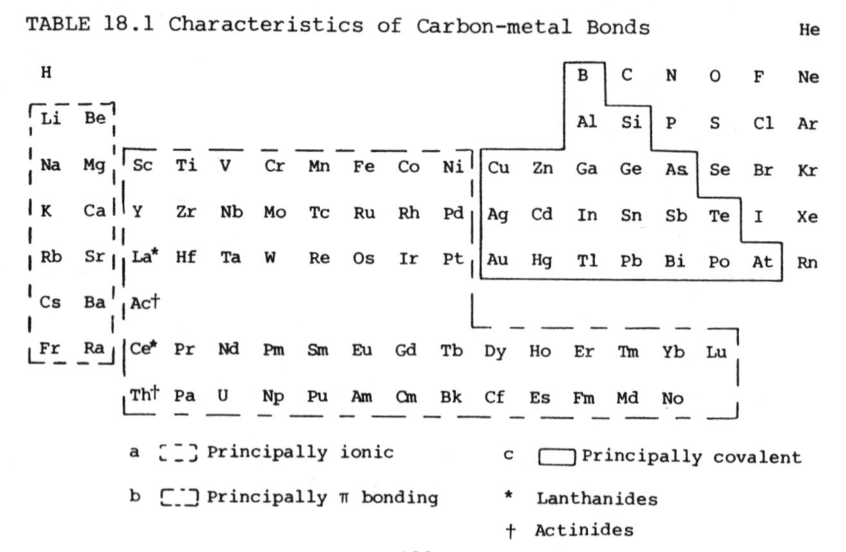

18.2 PROPERTIES OF ORGANOMETALLIC COMPOUNDS

A) Many organometallic compounds react vigorously with water or other protic compounds and with oxygen.

B) Many organometallic compounds decompose in water, but they are soluble in various inert aprotic organic solvents.

C) Methylsodium and methylpotassium are colorless amorphous solids.

D) Dimethylzinc is a colorless, mobile, strongly refractive liquid.

E) Dimethylmercury, trimethylaluminum, tetramethyltin and tetramethyllead are volatile liquids that distill without decomposition.

F) Other characteristics are given in Table 18.2

Table 18.2 Properties of Organometallic Compounds

Organometallic compound	Bond character	Physical properties	Reactive
Carbon-alkali metals Li<Na<K<Rb<Cs	Ionic Very polar	Saltlike Nonvolatile Insoluble in non-polar solvents	Highly re-active Inflammable in air Fast hydro-lysis
Carbon-earth alkali metals Be< Mg< Ca< Sr <Ba	↓	↓	↓
Carbon transition metals e.g., Cu >Ag >Au; Zn > Cd > Hg	Covalent Weakly polar	Volatile Soluble in non-polar solvents	Less reac-tive Stable in air Slow hydro-lysis

Table 18.2 continued

its reactivity is either decreased or increased as indicated in the following:
Substituents Y with (-) I and/or (-)M effect
1) Next to the carbon $\overset{\delta^{2\oplus}}{Me} \underset{\rightarrow}{:} \overset{\delta\ominus}{CH_2} \underset{\rightarrow}{:} \overset{\delta\ominus}{Y}$ Higher polarity of Me-C
2) On the metal $\overset{\delta^{2\ominus}}{Y} \underset{\leftarrow}{:} \overset{\delta\oplus}{Me} \underset{\leftarrow}{:} \overset{\delta\oplus}{CH_2} : R$ Smaller polarity of Me-C

18.3 PREPARATION OF ORGANOMETALLIC COMPOUNDS

By the action of an alkyl halide on zinc dust.

R–X + Zn, finely divided → R–Zn–X

By the action of alkyl halides on magnesium in dry ether to give organometallic halides.

R–X + Mg, dry ether → R–MgX (Grignard reagent)

By the action of alkyl magnesium or alkyl zinc halides on metallic halides of less active metals.

R–MgX/dry ether + HgX_2 → R–Hg–X + MgX_2

By the action of metallic sodium on dialkyl zinc or dialkyl mercury in dry benzene to give alkyl sodium.

Ex.

$(C_6H_5)_2Hg$ + 2Na, dry thiophene → $2C_6H_5$–Na + Hg

By the action of metallic sodium on diphenyl mercury in pure thiophene to given alkyl sodium.

$2C_6H_5$ + 2Na, dry thiophene → $2C_6H_5$–Na + Hg

By reaction between hydrocarbons and free metals.

$$C_6H_5-CH{=}CH-C_6H_5 + 2Na \longrightarrow C_6H_5-\ddot{C}H-\ddot{C}H-C_6H_5 \quad Na^{\oplus}\,{}^{\ominus}\,Na^{\oplus}$$

Stilbene

Brownish violet, "ionic-like"

By reaction of alkali metal and hydrocarbons which have acidic hydrogens.

$$C_6H_5-\underset{\ominus}{\ddot{C}H}-\underset{\ominus}{\ddot{C}H}-C_6H_5 \quad Na^{\oplus}\ Na^{\oplus}$$

Ex.

$$2(C_6H_5)_3C{-}H + 2K \rightarrow 2(C_6H_5)_3C{:}^{\ominus}K^{\oplus} + H_2$$

$$H{-}C \equiv C{-}H \xrightarrow[-H]{+Na} H{-}C \equiv C{:}^{\ominus}Na^{\oplus} \xrightarrow[+Na]{-H} Na^{+}{:}C^{-} \equiv C^{-}{:}Na^{+}$$

By the reaction between organic halides and metals.

$$RX \text{ or } C_6H_5X + \text{powdered metal} \xrightarrow[400°C]{200°-} \text{organometallic compound}$$

Powdered metals = Si, Al, Ge, Zn, Te, Sn.

Ex.

$$CH_3CH_2{-}Br + 2Li \xrightarrow{\text{in ether}} CH_3CH_2Li + LiBr$$

Halogen-metal interconversion

$$RX + R'Li \rightleftarrows RLi + R'X$$

RX = vinyl, alkyl, ethyarylbromides or iodides

R' = alkyl

$$Cl{-}C_6H_4{-}Br + CH_3CH_2CH_2CH_2Li \rightleftarrows Cl{-}C_6H_4{-}Li + CH_3(CH_2)_3Br$$

By the addition of a metal and hydrogen to alkenes.

Ex. $3(CH_3)_2C = CH_2 + Al + \frac{3}{2} H_2 \rightarrow [(CH_3)_2CH{-}CH_2]_3Al$

By metal-metal exchange

Ex. $(CH_3CH_2)_2Hg + 2Li \xrightarrow[65°, 3days]{Ligroin} 2CH_3CH_2Li + Hg$

By the reaction of metallic halides of less reactive metals with alkylmagnesium halides.

$4RMgX$, dry ether $+ 2PbCl_2 \rightarrow R_4Pb + Pb + 2MgX_2 + 2MgCl_2$

$2RMgX$, dry ether $+ HgCl_2 \rightarrow R_2Hg + MgX_2 + MgCl_2$

By the action of metallic halides of less reactive metals with dialkyl zinc.

R_2Zn, inert solvent $+ HgCl_2 \rightarrow R_2Hg + ZnCl_2$

By the action of intermetallic compounds with alkyl halides.

$2R{-}I + Na_2Hg \rightarrow R_2Hg + 2NaI$

$4R{-}X + 4NaPb \rightarrow R_4Pb + 4NaX + 3Pb$

Ex.

$4CH_3CH_2{-}Cl + 4Na{-}Pb \rightarrow (CH_3{-}CH_2)_4Pb + 4NaCl + 3Pb$

$2C_6H_5{-}Br + 2Na{-}Hg$, dry toluene $\rightarrow (C_6H_5)_2Hg + Hg + 2NaBr$

18.4 REACTIONS OF ORGANOMETALLIC COMPOUNDS

Grignard reagents react under proper conditions to give hydrocarbons when treated with reagents containing labile hydrogen atoms such as:

A) Acids

$R'{-}MgX + R{-}CO{-}OH \rightarrow R'{-}H + R{-}CO{-}O{-}MgX$

B) Alcohols

$R'{-}MgX + R{-}OH \rightarrow R'{-}H + R{-}O{-}MgX$

C) Thiols

$R'\text{-}MgX + R\text{-}SH \rightarrow R'\text{-}H + R\text{-}S\text{-}MgX$

D) Water

$R'\text{-}MgX + H_2O \rightarrow R'\text{-}H + HO\text{-}MgX$

E) Amines

$R'\text{-}MgX + R\text{—}NH_2 \rightarrow R'\text{-}H + R\text{-}NH\text{-}MgX$

F) Amides

$R'\text{-}MgX + R\text{-}CO\text{-}NH_2 \rightarrow R'\text{-}H + R\text{-}CO\text{-}NH\text{-}MgX$

Grignard reagents react under proper conditions to give condensation products when treated with reagents containing labile halogen atoms such as:

A) Phosgene

$2R'\text{-}MgX + Cl_2C{=}O \rightarrow R'_2C{=}O + MgX_2 + MgCl_2$

B) Alkyl halides

$R'\text{-}MgX + R\text{-}X \rightarrow R'\text{-}R + MgX_2$

C) Acid halides

$R'\text{-}MgX + R\text{-}OC\text{-}X \rightarrow R\text{-}C\text{-}O\text{-}R' \quad + MgX_2$

D) Aryl halides

$R'\text{-}MgX + Ar\text{-}X \rightarrow R'\text{-}Ar + MgX_2$

E) Halo derivatives

$R'\text{-}MgX + XCH_2\text{-}O\text{-}R \rightarrow MgX_2 + R\text{-}O\text{-}CH_2\text{-}R'$

Grignard reagents react under proper conditions to give addition products with reagents containing the carbonyl group.

R–	–C=O	–CH₂	–CHR	–CR₂	–C(OR)R	–C(NR₂)R
X–Mg–	–O	–O	–O	–O	–O	–O

Grignard reagents react to give addition products with reagents containing other active unsaturated linkages.

R–	–S=O	–C=S	–C–R	–C=N–R	–CH₂–CH₂	–C=N–Ar
X–Mg	–O	–S	‖ N⁻	|	–O	–O

Grignard reagents react under proper conditions to give double decomposition products with:

A) Organic anhydrides

$$R{-}MgX + R{-}\overset{\overset{\displaystyle O}{\|}}{C}{-}O{-}\overset{\overset{\displaystyle O}{\|}}{C}{-}R \rightarrow R{-}\overset{\overset{\displaystyle O}{\|}}{C}{-}R + R{-}\overset{\overset{\displaystyle O}{\|}}{C}{-}O{-}MgX$$

B) Sulfonic acid esters

$$R{-}MgX + ArSO_2OR \rightarrow 2R + Ar{-}SO_2{-}O{-}MgX$$

C) Inorganic halides
($AsCl_3$, $BiCl_3$, $GeCl_4$, $HgCl_2$, PCl_3, $SbCl_3$, $SiCl_4$)

$$3R{-}MgX + AsCl_3 \rightarrow 3MgX{-}Cl + R_3As$$

Organometallic compounds in which the metal has an electronegativity value of 1.7 or less react with water to give a hydrocarbon and a metal hydroxide.

Ex. $CH_3Li + H_2O \rightarrow CH_4 + LiOH$

$CH_3CH_2MgBr + H_2O \rightarrow CH_3CH_3 + HOMgBr$

$(CH_3)_3Al + 3H_2O \rightarrow 3CH_4 + Al(OH)_3$

Reactions with halogens

$RM + Cl_2 \rightarrow RCl + M^+Cl^-$ (vigorous)

Reactions with other organometallic compounds.

Alkyl copper compounds react with alkyl lithium compounds to give cuprates.

Ex. $CH_3Li + CH_3Cu \rightarrow (CH_3)_2CuLi$

18.5 USES OF ORGANOMETALLIC COMPOUNDS

As an important reagent for the synthesis of organic compounds (e.g., Grignard reagents).

Alkylboranes are important, and hydroboration is a method for preparing alcohols.

$$6RCH{=}CH_2 + (BH_3)_2 \rightarrow 2(RCH_2CH_2)_3B \rightarrow$$

$$\xrightarrow[6OH^-]{H_2O_2} 6RCH_2CH_2OH + 2B(OH)_3$$

Alkylmercury compounds are useful as intermediates in the formation of alcohols and ethers.

$$RCH{=}CH_2 + Hg(OCOCH_3)_2 \xrightarrow{H_2O} RCH(OH){-}CH_2HgOCOCH_3 \rightarrow$$

$$\xrightarrow{NaBH_4} RCH(OH)CH_3 + Hg$$

Organic compounds of cadmium and cuprates are often used for the preparation of ketones from acyl halides.

$$2R{-}\overset{O}{\overset{\|}{C}}{-}Cl + R'_2Cd \rightarrow 2R{-}\overset{O}{\overset{\|}{C}}{-}R' + CdCl_2$$

$$2R{-}\overset{O}{\overset{\|}{C}}{-}Cl + R'_2CuLi \rightarrow 2R{-}\overset{O}{\overset{\|}{C}}{-}R' + CuCl + LiCl$$

Mercuric compounds are toxic towards plant life and are used as bactericides, algicides, fungicides and herbicides.

Mercurochrome and methiolate are used medically.

18.6 GRIGNARD REAGENTS

Structural formula: $\overset{\delta\ominus}{R} - \overset{\delta\oplus}{MgX}$

R = Phenyl or Alkyl
X = halide

Preparation. Grignard reagents are prepared by the reaction between highly pure magnesium and an alkyl, or aryl halide.

Structure. Grignard reagents in ether exist in dimer form.

$$\begin{array}{ccc} R & & X \\ & Mg & \\ R & & X \end{array} \qquad Mg \cdots [O(C_2H_5)_2]_4$$

R.   .X
   Mg
R'   'X

$O(C_2H_5)_2$ (above), $O(C_2H_5)_2$, $O(C_2H_5)_2$, $O(C_2H_5)_2$ (below) coordinated to Mg

REACTIONS OF GRIGNARD REAGENTS

A) with compounds possessing an active hydrogen.

$$R{-}C\equiv CH + R'MgX \xrightarrow{-R'H} R{-}C\equiv C{-}MgX \xrightarrow{H_2O} R{-}C\equiv C{-}H$$

$$R{-}CH_2OH + R'MgX \xrightarrow{-R'H} R{-}CH_2{-}OMgX \xrightarrow{H_2O} R{-}CH_2OH$$

$$R_2NH + R'MgX \xrightarrow{-R'H} R_2N{-}MgX \xrightarrow{H_2O} R_2NH$$

B) with oxygen.

$$RMgX + O_2(\text{in ether}) \rightarrow R{-}O{-}O{-}MgX \xrightarrow{H_2O} R{-}O{-}O{-}H$$

hydroperoxides

C) with carbonyl compounds (addition).

$$R'R''C{=}O \xrightarrow{RMgX} R''{-}C(R')(R){-}OMgX \xrightarrow{H_2O} R''{-}C(R')(R){-}OH$$

(R', R'' = alkyl or H)

3° alcohol when R', and R" are alkyl

$$O{=}C{=}O \xrightarrow{RMgX} R{-}C({=}O)OMgX \xrightarrow{H_2O} R{-}COOH$$

carboxylic acid

$$R'(RO'')C{=}O \xrightarrow{\text{excess } RMgX} R{-}C(R')(R){-}OMgX \xrightarrow{H_2O} R{-}C(R')(R){-}OH$$

(R' = alkyl or H, R = alkyl)

3° alcohol when R' is alkyl

D) with α,β-unsaturated carbonyl compounds (addition).

$$\underset{\delta^{\oplus}}{R\text{-}CH}{=}CH\text{-}\overset{OR'}{\underset{\delta^{\ominus}}{C}}{=}O + R''MgX \rightarrow R\text{-}\overset{R''}{\underset{H}{C}}\text{-}\overset{H}{C}{=}\overset{OR'}{C}\text{-}OMgX \xrightarrow{H_2O}$$

1,4-Addition

$$R\text{-}\overset{R''}{\underset{H}{C}}\text{-}\overset{H}{C} = \overset{OR'}{C}\text{-}OH \rightarrow R\text{-}\overset{R''}{\underset{H}{C}}\text{-}\overset{H}{\underset{H}{C}}\text{-}\overset{OR'}{C} = O$$

an ester

E) with nitriles (addition).

$$R\text{-}C\equiv N + R'MgX \rightarrow R\text{-}\underset{R'}{C} = N\text{-}MgX \xrightarrow{H_2O} R\text{-}\underset{R'}{C} = NH$$

$$R\text{-}\overset{R'}{C} = O + NH_3 \xleftarrow{H_2O} R\text{-}\underset{R'}{C} = NH$$

ketone

F) with small cyclic ethers (ring opening reactions).

$$\text{(ethylene oxide)} + RMgX \longrightarrow R\text{-}CH_2CH_2OMgX \xrightarrow{H_2O}$$

$$\text{(oxetane)} + RMgX \longrightarrow R\text{-}CH_2CH_2CH_2OMgX \xrightarrow{H_2O}$$

} 1° Alcohols

G) with alkyl halides (displacement reactions); displacement of a halide ion, X^-, as in an alkyl chloride, by the nucleophile $R^{\delta\ominus}\text{-}Mg^{\delta\oplus}$.

$$CH_2 = CH\text{-}CH_2\text{-}Cl + R\text{-}MgX \xrightarrow{SN_2} CH_2 = CH\text{-}CH_2\text{-}R + MgXCl$$

CHAPTER 19

HETEROCYCLIC COMPOUNDS

19.1 STRUCTURE

Heterocyclics are cyclic compounds in which one or more of the ring atoms are not carbon. Heterocyclics include saturated and unsaturated cyclic ethers, thioethers and amines.

NOMENCLATURE

Number 1 is given to the heteroatom and numbering proceeds clockwise or counterclockwise so that the other substituents or other heteroatoms get the lowest numbers:

3-Methylpyrrole

When two or more different heteroatoms exist, oxygen takes preference over sulfur and nitrogen.

4-Methyloxazole

19.2 PROPERTIES OF FURAN, PYRROLE, AND THIOPHENE

Pyrrole   Furan   Thiophene

Furan, pyrrole and thiophene have the properties of unsaturated secondary amines, ethers, thioethers and conjugated dienes.

Furan undergoes Diels-Alder cycloaddition with maleic anhydride.

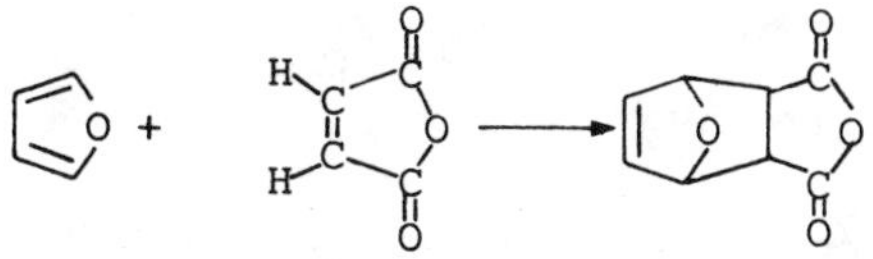

Furan, thiophene and pyrrole undergo electrophilic substitution in preference to addition reactions.

The three have been shown by microwave spectra to be planar molecules.

All three have aromatic characteristics.

19.3 SYNTHESIS OF FURAN, PYRROLE, AND THIOPHENE

GENERAL METHOD: THE PAAL-KNORR SYNTHESIS

Reaction occurs by heating 1,4-Diketones in the presence of a dehydrating agent; ammonia, or a sulfide to obtain furan, pyrrole or thiophene, respectively. Reaction proceeds by way of dienol.

1,4-Diketone ⇌ [Dienol] $\xrightarrow[P_2O_5]{-H_2O}$ 2,5-Dialkylfuran

$\xrightarrow[(NH_4)_2CO_3]{+NH_3,\ -2H_2O}$ 2,5-Dialkylpyrrole $\xrightarrow[P_2S_5]{+H_2S,\ -2H_2O}$ 2,5-Dialkylthiophene

KNORR PYRROLE SYNTHESIS

An α-amino ketone and a ketone, or a β-ketoester, give a substituted pyrrole.

$$\xrightarrow{-2H_2O}$$

2,4-Dimethyl-3,5-dicarbethoxy-pyrrole

Thiophene is prepared as follows:

$$n-C_4H_{10} + S \xrightarrow{600°C} \text{thiophene} + H_2S$$

19.4 REACTIONS

Electrophilic substitution in furan is at the α position.

a) Furan + Acetic anhydride ($CH_3COOCCH_3$) $\xrightarrow{BF_3}$ 2-Acetylfuran (75-92%)

b) Furan + Br_2 $\xrightarrow[25°C]{dioxane}$ (75%) Bromofuran

c)

$$\text{2-Nitrofuran} \xrightarrow{HNO_3} \text{2,5-Dinitrofuran}$$

2-Nitrofuran 2,5-Dinitrofuran

Electrophilic substitution in thiophene is at the α position.

a) Thiophene + CH_3CONO_2 (Acetyl-nitrate) $\xrightarrow[-10°C]{Ac_2O}$ 2-Nitrothiophene (70%) + 3-Nitrothiophene (5%)

Thiophene Acetyl-nitrate 2-Nitroth-iophene (70%) 3-Nitrothi-ophene (5%)

b) Thiophene + CH_3COCCH_3 (Acetic anhydride) $\xrightarrow{H_3PO_4}$ 2-Acetylthiophene ($COCH_3$) (94%)

Thiophene Acetic anhy-dride (94%) 2-Acetylthiophene

c) Thiophene + I_2 $\xrightarrow[\text{benzene}]{HgO}$ 2-Iodothiophene

Thiophene (75%) 2-Iodothiophene

d) Methyl 5-methylthiophene 2-Carboxylate $\xrightarrow[ZnCl_2, CHCl_3]{CH_2O, HCl}$ Methyl-5-Methyl-4-chloromethyl-2-carboxylate ($ClCH_2$, H_3C, CO_2CH_3) (93%)

Methyl 5-methylthiophene 2-Carboxylate Methyl-5-Methyl-4-chloro methyl-2-carboxylate

e) 2-Methylthiophene $\xrightarrow[Ac_2O]{HNO_3}$ 5-Nitro-2-methylthiophene (70%) + 3-Nitro-2-methylthiophene (30%)

2-Methylthiophene (70%) 5-Nitro-2-meth-ylthiophene (30%) 3-Nitro-2-methylthiophene

Electrophilic substitution in pyrroles at the α position.

Pyrrole + CH_3COCCH_3 (Acetic anhydride) $\longrightarrow$ 2-Acetylpyrrole ($COCH_3$) (60%)

Pyrrole (60%) 2-Acetylpyrrole

Polymerization of pyrroles by dilute acids.

$$\text{Pyrrole} \underset{}{\overset{H^{\oplus}}{\rightleftharpoons}} \text{2H-pyrrolium } (\oplus NH) + \text{pyrrole} \longrightarrow \text{dimer } (N^{\oplus}H)$$

Electrophilic substitution at α position.

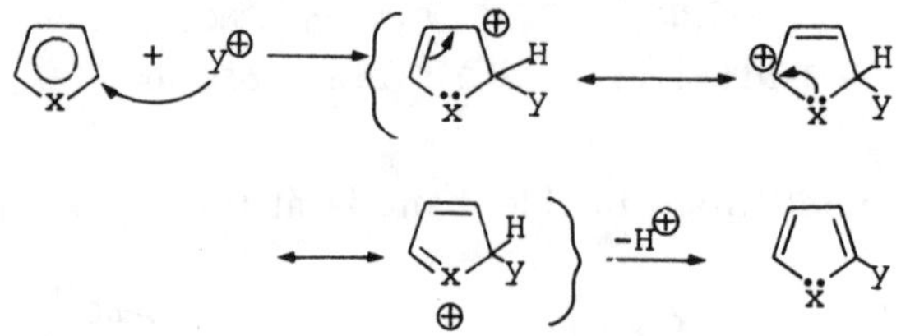

Acidic character of pyrroles.

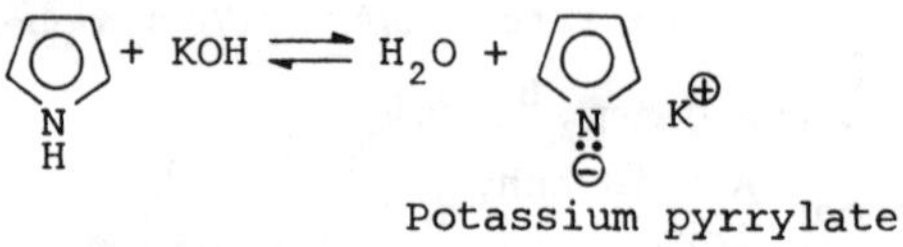

Potassium pyrrylate

Pyrrole is a weak acid which yields alkali-metal salts upon treatment with alkaline hydroxides.

19.5 PYRIDINE, QUINOLINE, AND ISOQUINOLINE

STRUCTURE

The replacement of one CH group in benzene by nitrogen yields the pyridine molecule. It is a hybrid of the two Kekulé structures and the three polar structures shown below.

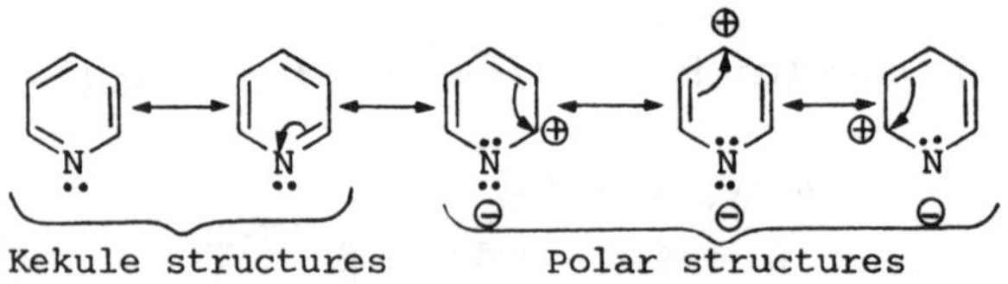

Kekule structures   Polar structures

19.6 STRUCTURAL PROPERTIES OF PYRIDINES

Pyridine undergoes both nucleophilic substitution and electrophilic substitution.

The basic properties of pyridine (which are similar to those of the tertiary amines), are not influenced by the cyclic delocalization of the π electrons.

19.7 SYNTHESIS OF PYRIDINE, QUINOLINE, AND ISOQUINOLINE

1,4 Cycloaddition of nitriles to 1,3 dienes in the presence of an oxidizing agent produces 2-alkyl pyridine.

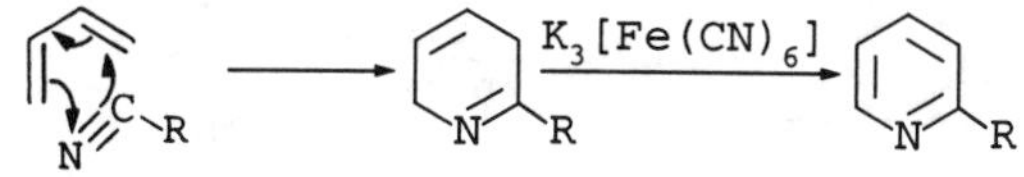

$K_3[Fe(CN)_6]$

The oxidation of 1,2-dihydroquinoline by nitrobenzene yields quinoline (Skraup quinoline synthesis).

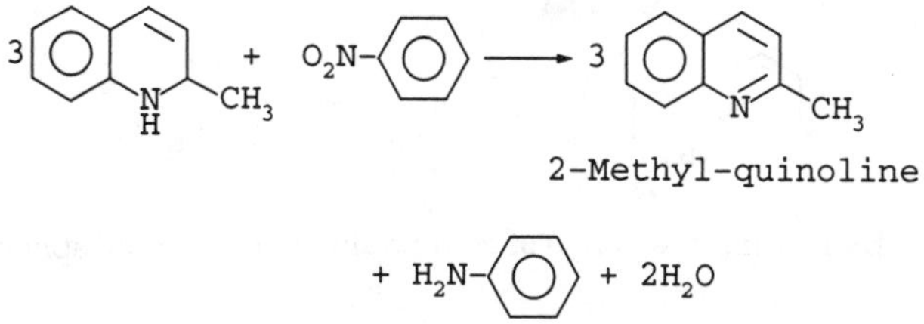

2-Methyl-quinoline

The cyclization of N-acyl-β-phenylethyl amines in the presence of phosphorus pentoxide as a dehydration catalyst yields isoquinolines (Bischler-Napieralski synthesis).

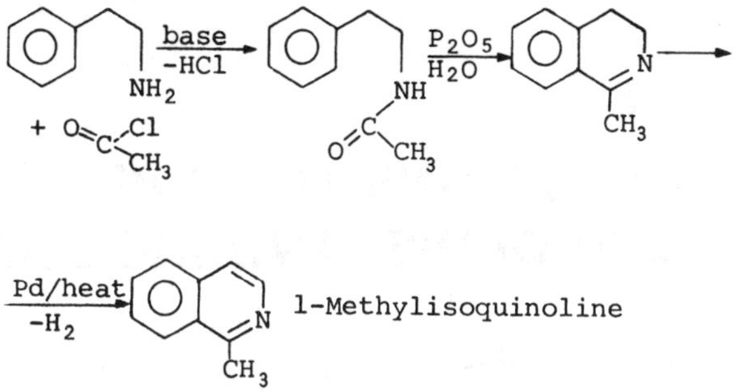

1-Methylisoquinoline

The condensation of o-aminobenzaldehyde with a ketone (Friedländer synthesis).

$$\text{o-aminobenzaldehyde} + CH_3\overset{O}{\overset{\|}{C}}CH_3 \xrightarrow{\text{pH 12}} \text{2-methylquinoline} \ (85\%)$$

$$\text{o-aminobenzaldehyde (CHO, NH}_2\text{)} + \text{cyclohexanone} \xrightarrow[\text{EtOH}]{\text{NaOH}} \text{1,2,3,4-tetrahydroacridine}$$

Substitution of α, ß–unsaturated ketone or aldehyde.

$$\text{aniline (NH}_2\text{)} + CH_2{=}CH\overset{O}{\overset{\|}{C}}CH_3 \xrightarrow[ZnCl_2]{FeCl_3} \text{4-methylquinoline}$$

(73%) Lepidine
(4-methyl-quinoline)

Heating a mixture of aniline, glycerol, nitrobenzene and sulfuric acid (Skraup synthesis) to obtain quinoline.

$$\text{aniline (NH}_2\text{)} + \begin{matrix} CHO \\ CH \\ \| \\ CH_2 \end{matrix} \longrightarrow C_6H_5\text{–NH–}CH_2CH_2CHO \longrightarrow \text{1,2-dihydroquinoline} \xrightarrow{[O]} \text{quinoline}$$

By heating the oxime of cinnamaldehyde with phosphorus pentoxide.

$$C_6H_5CH{=}CH\text{–}CH{=}NOH \xrightarrow[\text{heat}]{P_2O_5} \text{Quinoline (positions 1–8; N = 1)}$$

Quinoline

19.8 REACTIONS OF PYRIDINE, QUINOLINE, AND ISOQUINOLINE

Methylation of Pyridine

$$\text{Pyridine} + CH_3I \longrightarrow \text{N-Methylpyridinium iodide}$$

1° Alkyl halide

N-Methylpyridinium-iodide

Sulfonation of Pyridine with Sulfur Trioxide.

$$\text{Pyridine} + SO_3 \longrightarrow \text{N-Pyridine sulfite}$$

N-Pyridine sulfite

Reduction of Pyridine to Piperidine.

$$\text{Pyridine} + 3H_2 \xrightarrow[\text{3atm.}]{\text{Pd, 25°C}} \text{Piperidine}$$

Bromination reaction.

$$\text{Pyridine} \xrightarrow[\text{200-300°C, Pressure}]{+Br_2, -HBr} \text{3-Bromopyridine and 3,5-Dibromopyridine}$$

3-Bromopyridine 3,5-Dibromo-pyridine

Oxidation of Pyridine to Yield Pyridine N-oxide.

$$\text{Pyridine} + H_2O_2 \xrightarrow{-H_2O} \text{Pyridine N-oxide (resonance structures)}$$

Amination of Pyridines to Give 2-Aminopyridine

$$\text{Pyridine} + :NH_2^{\ominus} \xrightarrow{NaNH_2} \text{(resonance intermediates)} \xrightarrow{-:H^{\ominus}} \text{2-Aminopyridine}$$

2-Aminopyridine

Nitration of Pyridine–N–Oxide to yield 4–Nitropyridine–N–Oxide

$$\text{Pyridine-N-oxide} + NO_2^{\oplus} \xrightarrow{HNO_3} [\text{4-H,4-NO}_2\text{ adduct}] \xrightarrow{-H^{\oplus}} \text{4-Nitropyridine-N-oxide}$$

Arylation and alkylation of pyridines with lithium aryls and alkyls, respectively, yield the corresponding 2–aryl and 2–alkylpyridines.

$$\text{Pyridine} + Li^{\delta\oplus}\text{-}C_6H_5^{\delta\ominus} \longrightarrow [\text{adduct}]\,Li^{\oplus} \xrightarrow{-Li^{\oplus}H^{\ominus}} \text{2-Phenylpyridine}$$

2-Phenylpyridine

Electrophilic substitution reactions.

$$\text{Quinoline} \xrightarrow[0°C,\ 30\ \text{Mins.}]{HNO_3,\ H_2SO_4} \text{5-Nitroquinoline (52\%)} + \text{8-Nitroquinoline (48\%)}$$

(52%) 5-Nitroquinoline

(48%) 8-Nitroquinoline

$$\text{Isoquinoline} \xrightarrow[0°C,\ 30\ \text{Mins.}]{HNO_3,\ H_2SO_4} \text{5-Nitroisoquinoline (90\%)} + \text{8-Nitroisoquinoline (10\%)}$$

(90%) 5-Nitroisoquinoline

(10%) 8-Nitroisoquinoline

Nucleophilic substitution reactions.

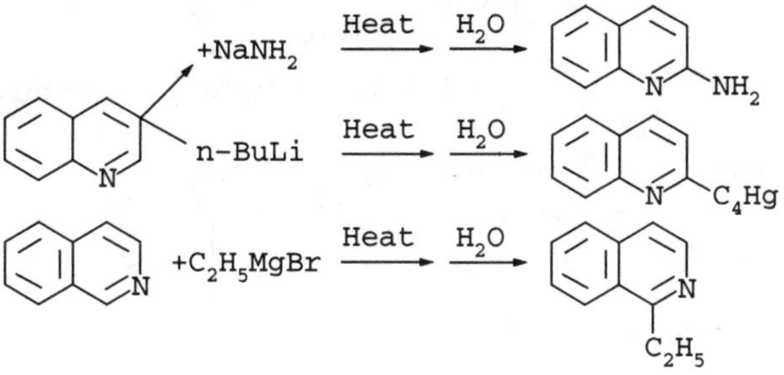

Quinoline $\xrightarrow{+NaNH_2}$ $\xrightarrow{\text{Heat}}$ $\xrightarrow{H_2O}$ 2-aminoquinoline (NH_2)

Quinoline $\xrightarrow{\text{n-BuLi}}$ $\xrightarrow{\text{Heat}}$ $\xrightarrow{H_2O}$ 2-substituted quinoline (C_4Hg)

Isoquinoline $+C_2H_5MgBr$ $\xrightarrow{\text{Heat}}$ $\xrightarrow{H_2O}$ 1-ethylisoquinoline (C_2H_5)

19.9 CONDENSED FURANS, PYRROLES, AND THIOPHENES

STRUCTURE

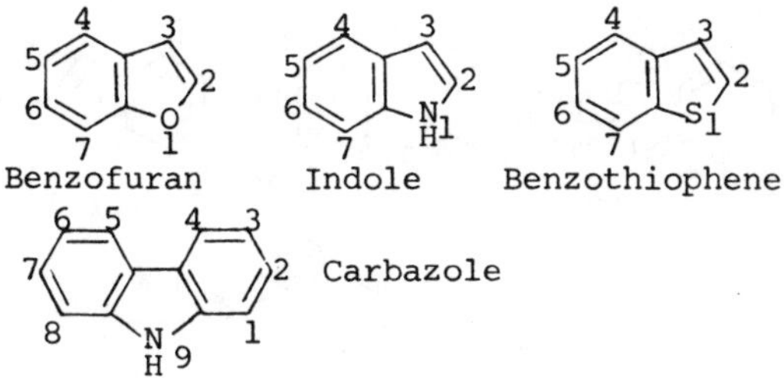

Benzofuran Indole Benzothiophene

Carbazole

The rings are numbered starting with the heteroatom, except for carbazole.

Synthesis

A) Fisher indole synthesis. The phenyl hydrazone of an aldehyde or ketone is treated with a catalyst such as BF_3, $ZnCl_2$ or polyphosphoric acid.

N N H → BF$_3$ / CH_3COOH / 65°C → N H (93%)

1,2,3,4-Tetrahydrocarbozole

N N H CH_3 → *PPA / 100°C → N H (73%)

*PPA=polyphosphoric acid

2-phenylindole

B) Phenylhydrazone of pyruvic acid reacts to yield indole-2-carboxylic acid, which can be decarboxylated to produce indole.

H_3C COOH N N H → PCl_5 → N H COOH → $-CO_2$ / 250° C → N H

C) Benzofuran is prepared from coumarin, which is itself prepared form salicylaldehyde by the Perkin synthesis.

O O → Br_2 (70%) → Br Br O O → KOH / EtOH (85%) → O -COOH

Coumarin

Coumarillic acid

O

Benzofuran

19.10 ALKALOIDS

Alkaloids are compounds of vegetable origin with heterocyclic ring system containing one or more basic nitrogen atoms. Most alkaloids are optically active and have a variety of structures. They are toxic and some act as narcotics.

Coniine

Nicotine

Quinine, an antimalarial

R=H :Morphine
R=CH_3 : Codeine
Narcotics

19.11 PYRIMIDINES AND PURINES

STRUCTURE

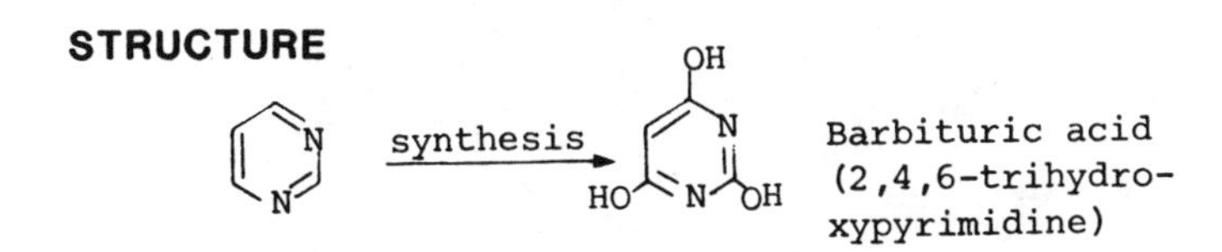

Derivatives of pyrimidine based on 2,4,6-trihydroxypyrimidine, are called barbituric acids. The above structure is only one of several tautomeric forms.

Barbituric acids can be synthesized from urea and alkyl- or aryl-substituted diethyl malonate.

$-2C_2H_5OH$

Phenobarbital

$R_1 = C_6H_5$
$R_2 = C_2H_5$

Adenine, a derivative of purine, [purine structure] is synthesized by the cyclization of formamidine and phenyl azo malodinitrine by way of 4,5,6-triaminopyrimidine.

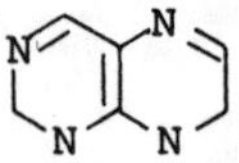

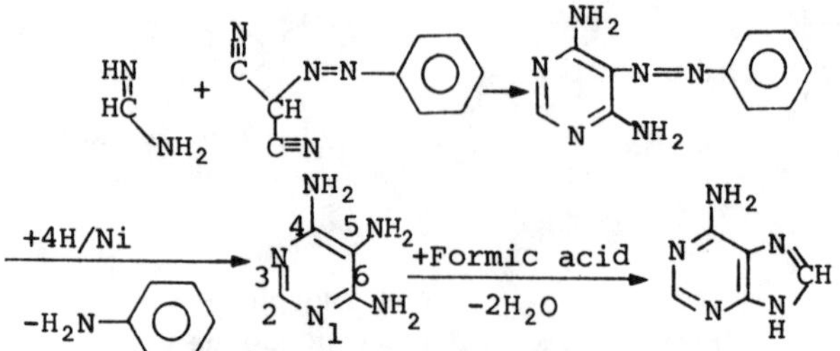

4,5,6-Triamino-pyrimidine

Adenine

+4H/Ni

$-H_2N-C_6H_5$

+Formic acid

$-2H_2O$

Uric acid is 2,4,6-trihydroxypurine, it exists in several tautomeric lactam forms: caffeine, the alkaloid of the coffee bean and tea plant, is 3,5,7-tri-N-methyl-4,6-dioxopurine.

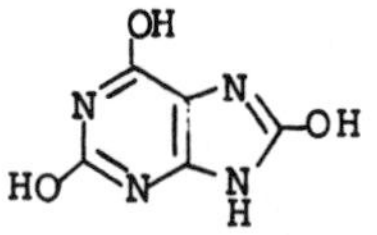

Uric acid

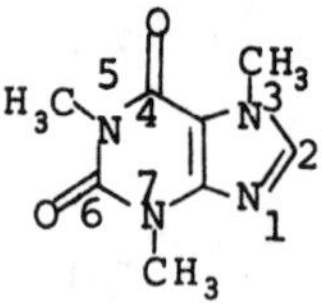

Caffeine

19.12 AZOLES

STRUCTURES AND NOMENCLATURES

Azoles are five-membered-ring aromatic heterocycles containing two nitrogens, one nitrogen and one oxygen or one nitrogen and one sulfur.

Thiazole

Pyrazole

Oxazole

Synthesis of Pyrazoles and Isoxazoles:

A)

$CH_3COCH_2COCH_3$ → (H_2NOH, HCl; H_2O, heat) → 3,5-Dimethylisoxazole (85%)

$CH_3COCH_2COCH_3$ → (H_2NNH_2, NaOH; H_2O, 15°) → 3,5-Dimethylpyrazole (73-77%)

The above reaction occurs by the action of hydrazine or hydroxylamine with 1,3-dicarbonyl compound or its equivalent.

B) Synthesis of isoxazoles can also occur by the cycloaddition of a nitrile oxide to an acetylene.

$C_6H_5C{\equiv}N^{\oplus}{-}O^{\ominus}$ + $C_6H_5C \equiv CCOOH$ → 3,4-Diphenylisoxazole-5-carboxylic acid

Benzonitrile oxide

C) General synthesis of 1,3-azoles involves the dehydration of 1,4-dicarbonyl compounds (a form of Paal-Knorr cyclization).

$C_6H_5CO{-}NHCH_2COC_6H_5$ → (H_2SO_4, heat) → 2,5-Diphenyloxazole

$C_6H_5CO{-}NHCH(C_6H_5)COC_6H_5$ → ($NH_4^{\oplus}OAc^{\ominus}$, HOAc 120°C) → 2,4,5-Triphenylimidazole (93%)

$CH_3COCH_2{-}NH{-}COCH_3$ + P_2S_5 → (120°C) → 2,5-Dimethylthiazole

REACTIONS OF AZOLES

The azoles are significantly less reactive than furan, pyrrole and thiophene. Their order of reactivity of 1,2-azoles is

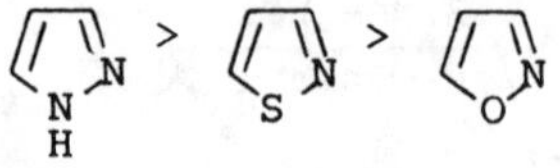

A) Electrophilic substitution reaction.

HNO_3, H_2SO_4, 115°C, 19hrs.

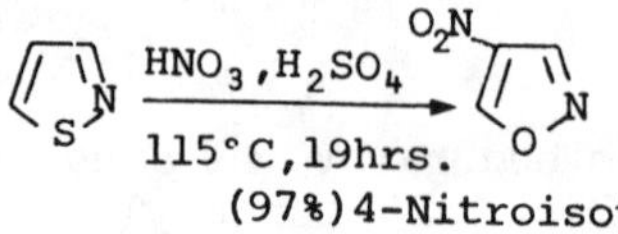

(97%) 4-Nitroisothiazole

B) Substitution reaction for imidazoles.

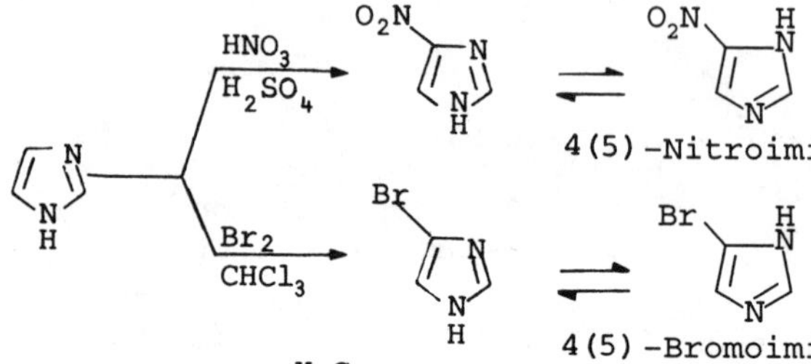

HNO_3 / H_2SO_4

4(5)-Nitroimidazole

Br_2 / $CHCl_3$

4(5)-Bromoimidazole

Br_2 / $CHCl_3$

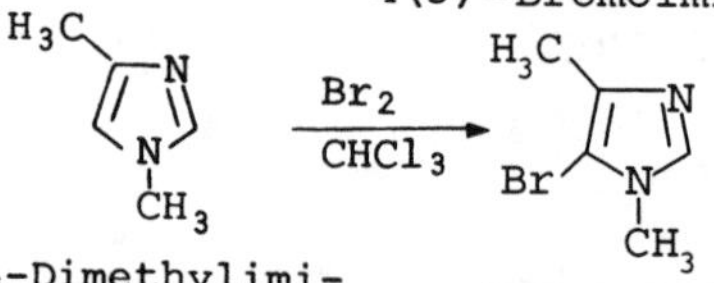

1,4-Dimethylimidazole

5-Bromo-1,4-dimethylimidazole

CHAPTER 20

CARBOHYDRATES

Simple carbohydrates are polyhydroxy aldehydes and polyhydroxy ketones existing in cyclic hemiacetal and hemiketal forms. They are, actually or potentially, hydroxy or polyhydroxy oxoderivatives of the hydrocarbons.

Structural formula: $n[C_x(H_2O)_x] - (n - 1)H_2O$

x = number of carbon atoms in a building unit.

n = number of building units per molecule.

20.1 NOMENCLATURE AND CLASSIFICATION

Carbohydrates with an aldehyde group are called aldoses and those with a keto group are called ketoses.

```
  CHO                  CH2OH
   |                    |
 (CHOH)n                C = O
   |                    |
  CH2OH               (CHOH)n
                        |
                       CH2OH
```

An aldose (Polyhydroxy aldehyde)

A ketose (Polyhydroxy ketone)

Aldoses and ketoses are further characterized as aldo- or keto-, trioses, -tetroses, -pentoses, -hexoses, etc., according to the number of carbon atoms they have.

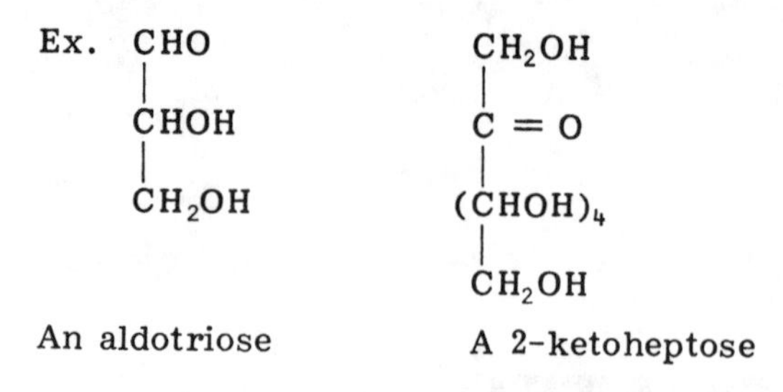

Carbohydrates are classified as:

A) Monosaccharides: single aldoses and ketoses which do not hydrolyze to simpler sugars.

B) Oligosaccharides: di-, tri-, tetra-, etc., saccharides which can be hydrolyzed to monosaccharides (up to 10).

C) Polysaccharides: macromolecule consisting of higher numbers (up to thousands) of monosaccharides.

Saccharides containing a non-sugar component are called glycosides.

20.2 PHYSICAL PROPERTIES

A) Monosaccharides are readily soluble in water, have a sweet taste, which increases with the number of -OH groups, and are usually crystalline.

B) Glycolic aldehyde is sweet and soluble in cold water and alcohol, but insoluble in ether. It can be obtained in crystalline form (m.p. 95-97°C).

C) Glyceraldehyde (m.p. 138°C) is a white powder, not very sweet, soluble in water, slightly soluble in alcohol and insoluble in ether.

D) Dihydroxyacetone is sweet, very soluble in cold water, slightly soluble in ether, almost insoluble in hot acetone and insoluble in ligroin.

E) Arabinose is precipitated by alcohol and yields an orange yellow osazone.

F) D-fructose forms anhydrous crystals which are somewhat hygroscopic. It forms an insoluble methylphenylhydrazone whereas the corresponding derivative of D-glucose is soluble.

G) D-Mannose may be precipitated by alcohol.

H) Most of the di- and trisaccharides are soluble in water and crystallize quite readily.

I) The sweetness of sucrose is taken as 100 (standard basis). The relative sweetness of common sugars is indicated as:

D-fructose......173.3	Rhamnose 32.5
D-glycose........74.3	D-galactose.....32.1
D-xylose 40.0	Raffinose.......22.6
Maltose........32.5	Lactose.........16.0

J) Polysaccharides, as a rule are not soluble in water and are not readily obtained in a crystalline form.

K) Dextrin has an insipid taste and gives a red to brown coloration.

L) Glycogen tends to give a milky solution which may be clarified by treating with acetic acid. With iodine in potassium iodide, glycogen gives a reddish-brown coloration.

M) Ordinary air-dried starch is tasteless. It contains water but becomes anhydrous on heating to 110°C.

N) On boiling starch with water, the granules break to give a colloidal solution which gives a blue coloration with iodide in potassium iodide. Coloration disappears upon heating and reappears upon cooling.

O) Cellulose may be fibrous, cellular or woody.

20.3 PREPARATION OF MONOSACCHARIDES

Aldol condensation

A) X–H–CHO/lime water → $CH_2OH–CHO$, $CH_2OH–CHOH–CHO$,

$CH_2OH–CHOH–CHOH–CHO$, etc.

B) $2CH_2OH–CHO$/lime water → $CH_2OH–CHOH–CHOH–CHO$

C) X–$CH_2OH–HCOH–CHO$ ⇄ y$CH_2OH–CO–CH_2OH$/lime water

→ $CH_2OH–HCOH–HCOH–HCOH–CO–CH_2OH$

KILIANI–FISCHER SYNTHESIS

Ra is $CH_2OH–HCOH–HCOH–HCOH–$

$$2\,\mathrm{H{-}\underset{\underset{Ra}{|}}{C}{=}O} \xrightarrow[\text{trace } NH_3]{3\%\ HCN} \mathrm{H{-}\underset{\underset{Ra}{|}}{\overset{\overset{CN}{|}}{C}}{-}OH} + \mathrm{HO{-}\underset{\underset{Ra}{|}}{\overset{\overset{CN}{|}}{C}}{-}H}$$

$$\xrightarrow[\text{dil. } H_2SO_4\text{, boil}]{\text{hydrolysis}} \mathrm{H{-}\underset{\underset{Ra}{|}}{\overset{\overset{CO{-}OH}{|}}{C}}{-}OH} + \mathrm{HO{-}\underset{\underset{Ra}{|}}{\overset{\overset{CO{-}OH}{|}}{C}}{-}H}$$

After the polyhydroxy monoacids are isolated in the form of their lactones, then

$$\begin{array}{c} O{=}C\text{ (ring to O below)} \\ | \\ H{-}C{-}OH \\ | \\ HO{-}C{-}H \\ | \\ H{-}C{-}O \\ | \\ H{-}C{-}OH \\ | \\ CH_2OH \end{array} + 2Na/(Hg)/2H_2SO_4\text{,aq} \rightarrow \begin{array}{c} O{=}C{-}H \\ | \\ H{-}C{-}OH \\ | \\ HO{-}C{-}H \\ | \\ H{-}C{-}OH \\ | \\ H{-}C{-}OH \\ | \\ CH_2OH \end{array} + 2NaHSO_4$$

gluconolactone D-glucose

Mild oxidation, or more vigorous oxidation followed by reduction of the polyhydroxy alcohols.

A) Direct oxidation of corresponding alcohols.

$$\underset{\text{Ra}}{\overset{|}{\text{CH}_2\text{OH}}} + \text{Br}_2\text{, aq}/2\text{C}_6\text{H}_5\text{CO–ONa} \rightarrow \underset{\text{Ra}}{\overset{|}{\text{H–C=O}}} + 2\text{NaBr} + 2\text{C}_6\text{H}_5\text{–CO–OH}$$

B) Oxidation of a polyhydroxy alcohol to the corresponding monoacid and subsequent reduction of the monoacid lactone to the corresponding aldose.

$$\underset{\text{Rg}}{\overset{|}{\text{CH}_2\text{OH}}} + 2\text{Br}_2/\text{H}_2\text{O}/4\text{C}_6\text{H}_5\text{CO–ONa, aq} \rightarrow \underset{\text{Rg}}{\overset{|}{\text{O=C–OH}}} + 4\text{NaBr} + 4\text{C}_6\text{H}_5\text{–CO–OH}$$

Rg is $CH_2OH–HCOH–HCOH–HOCH–HCOH–$

Degradation reaction. D-glucose may be degraded to D-arabinose.

H–C=O / H–C–OH / HO–C–H / H–C–OH / H–C–OH / CH_2OH $\xrightarrow{H_2NOH}$ H–C=N–OH / H–C–OH / HO–C–H / H–C–OH / H–C–OH / CH_2OH $\xrightarrow[AcONa]{Ac_2O}$ C≡N / H–C–OAc / AcO–C–H / H–C–OAc / H–C–OAc / CH_2OAc

$\xrightarrow{\text{Alc. ammoniacal silver nitrate}}$ C≡N / H–C–OH / HO–C–H / H–C–OH / H–C–OH / CH_2OH $\xrightarrow{\text{Acetamide}}$ H–C $(NHAc)_2$ / HO–C–H / H–C–OH / H–C–OH / CH_2OH

$$\xrightarrow{\text{Dil. HCl}} \begin{array}{c} \text{H-C=O} \\ | \\ \text{HO-C-H} \\ | \\ \text{H-C-OH} \\ | \\ \text{H-C-OH} \\ | \\ \text{CH}_2\text{OH} \end{array}$$

Hydrolysis of polysaccharides

Ex. Arabinosans yield Arabinose
Xylosans yield Xylose
Dextrosans yield Glucose
Mannosans yield Mannose

Starch is hydrolyzed commercially to give a variety of products, such as corn syrup, grape sugar and dextrin, by controlling the extent of the hydrolysis.

By the condensation of aldoses with nitromethane, conversion to the salt with alkali, and subsequent hydrolysis with acid.

20.4 PREPARATION OF DISACCHARIDES

Isolation from natural products

A) Sucrose from sugar cane or sugar beet.

B) Maltose from sprouted barley or by partial hydrolysis of starch.

C) Lactose, as by-product in the cheese industry.

20.5 PREPARATION OF POLYSACCHARIDES

A) Dextrin is obtained by the controlled hydrolysis of starch.

B) Glycogen is obtained by the extraction of liver tissue.

C) Starch is produced from corn, wheat, potatoes, rice, arrow root and the sago palm.

20.6 REACTIONS OF CARBOHYDRATES

Osazone formation by reaction between α or β-D-glucose and 3 moles of phenylhydrazine.

α-D-Glucose + 3 Phenylhydrazine (C_6H_5-$NHNH_2$) $\xrightarrow[H_2O]{20°C}$ Glucosazone $+C_6H_5\text{-}NH_2+NH_3+2H_2O$

Glucosazone: H-C=N-NH-C_6H_5 / C=N-NH-C_6H_5 / HO-C-H / H-C-OH / H-C-OH / CH_2OH

Reduction. Treatment of aldoses with Na/Hg in alkaline solution is by catalytic hydrogenation.

D-Glucose $\xrightarrow[OH^{\ominus} \text{ in } H_2O]{Na/Hg}$ D-Sorbitol (optically active)

D-Sorbitol: CH_2OH / H-C-OH / HO-C-H / H-C-OH / H-C-OH / CH_2OH

Oxidation. Treatment of aldoses with nitric acid to cause the conversion to polyhydroxy–α, ω–dicarboxylic acids.

Ex. D-Erythrose (active): CHO / H—C—OH / H—C—OH / CH_2OH $\xrightarrow[0°C]{HNO_3, NaNO_2}$ meso-Tartaric acid (inactive): COOH / H—C—OH / H—C—OH / COOH

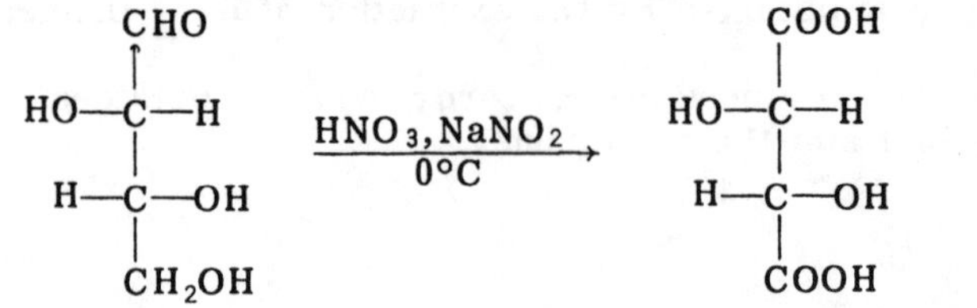

$$\begin{array}{c} CHO \\ HO-C-H \\ H-C-OH \\ CH_2OH \end{array} \xrightarrow[0°C]{HNO_3, NaNO_2} \begin{array}{c} COOH \\ HO-C-H \\ H-C-OH \\ COOH \end{array}$$

D-Threose (active) D-Tartaric acid (active)

KILIANI–FISCHER SYNTHESIS

The conversion of an aldose to two epimers of the next higher aldose by the application of the cyanohydrin formation of aldehydes.

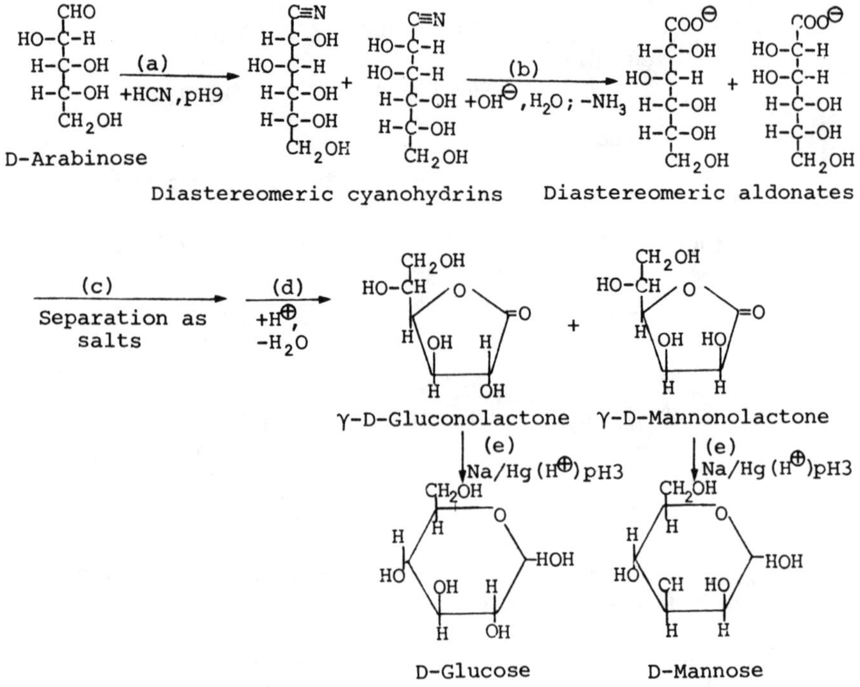

$$\begin{array}{c} CHO \\ HO-C-H \\ H-C-OH \\ H-C-OH \\ CH_2OH \end{array} \xrightarrow[+HCN, pH9]{(a)} \begin{array}{c} C\equiv N \\ H-C-OH \\ HO-C-H \\ H-C-OH \\ H-C-OH \\ CH_2OH \end{array} + \begin{array}{c} C\equiv N \\ HO-C-H \\ HO-C-H \\ H-C-OH \\ H-C-OH \\ CH_2OH \end{array} \xrightarrow[+OH^{\ominus}, H_2O; -NH_3]{(b)} \begin{array}{c} COO^{\ominus} \\ H-C-OH \\ HO-C-H \\ H-C-OH \\ H-C-OH \\ CH_2OH \end{array} + \begin{array}{c} COO^{\ominus} \\ HO-C-H \\ HO-C-H \\ H-C-OH \\ H-C-OH \\ CH_2OH \end{array}$$

D-Arabinose

Diastereomeric cyanohydrins Diastereomeric aldonates

$\xrightarrow{(c)}$ Separation as salts $\xrightarrow{(d)}$ $+H^{\oplus}$, $-H_2O$

γ-D-Gluconolactone + γ-D-Mannonolactone

(e) Na/Hg ($H^{\oplus}$) pH3

D-Glucose D-Mannose

Formation of glycosides by boiling aldoses in alcohol in the presence of anhydrous hydrogen chloride.

CH2OH, O, H, H, HOH, HO, OH, HO, H, H — D-Mannose

\+ CH_3OH $\xrightarrow[\text{heat}]{HCl}$

CH2OH, O, H, H, H, HO, OH, HO, OCH_3, H, H + CH2OH, O, H, H, OCH_3, HO, OH, HO, H, H, H

D-Mannose — Methyl-D-manopyranosides

Upon esterification by treatment with acetic anhydride/$ZnCl_2$ or with acetyl chloride, aldohexoses yield diastereomeric α- and β-glycopyranose pentaacetates.

CH_2OH, O, H, H, HOH, HO, OH, H, H, OH — D-Glucose

$\xrightarrow[-5CH_3COOH]{+5(CH_3CO)_2O, ZnCl_2}$

CH_2OCOCH_3, O, H, H, H, CH_3COO, $OCOCH_3$, $OCOCH_3$, H, H, $OCOCH_3$ + CH_2OCOCH_3, O, H, H, $OCOCH_3$, CH_3COO, $OCOCH_3$, H, H, H, $OCOCH_3$

D-Glucose — α-D-Glucopyranose pentaacetate — β-D-Glucopyranose pentaacetate

Methylation. Polyhydroxy compounds can be converted to ether groups.

CH_2OH, O, H, H, HOH, HO, OH, H, H, HO — D-Glucose

$\xrightarrow[-Na_2SO_4, H_2O]{(CH_3)_2SO_4, NaOH}$

CH_2OCH_3, O, H, H, H, H_3CO, OCH_3, H, OCH_3, H, OCH_3 + CH_2OCH_3, O, H, H, OCH_3, H_3CO, OCH_3, H, H, H, OCH_3

D-Glucose — Methyl-α(and β)-2,3,4,6-tetra-O-methyl-D-glycopyranoside(a permethylated sugar)

REACTION WITH ALKALI AND MINERAL ACIDS

By the treatment of glucose with calcium hydroxide in alkaline medium, many products are obtained. This is due to isomerization of α-ketols.

H—C=O / H—C—OH — α-ketol

H—C—OH ═ C—OH — enediol

⇌ CH_2—OH / C = O — β-ketol

⇌ H—C=O / HO—C—H — α-ketol

Treatment of aldoses with strong mineral acids causes dehydration, yielding furfural in the case of pentoses. Hexoses yield 5-hydroxymethyl furfural.

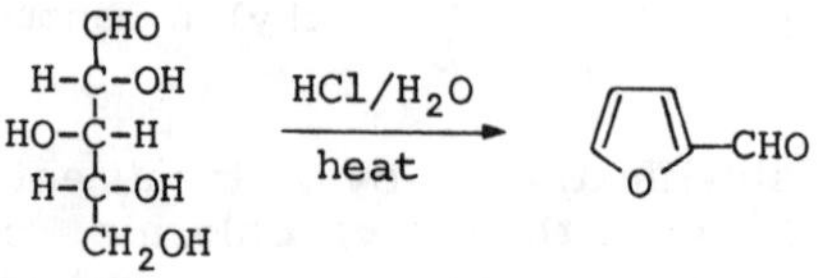

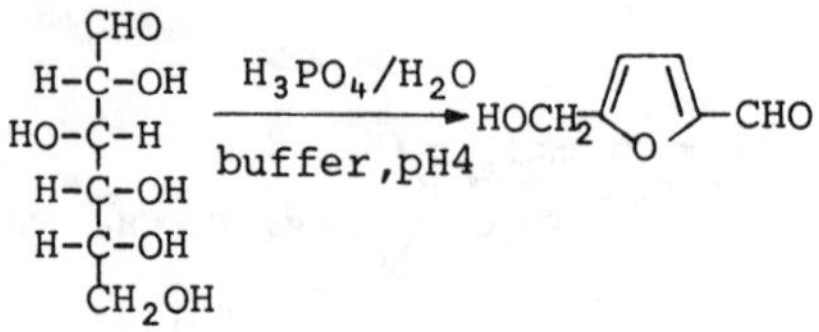

CHO
H-C-OH
HO-C-H
H-C-OH
CH_2OH
$\xrightarrow[\text{heat}]{HCl/H_2O}$ furan-CHO (O)

CHO
H-C-OH
HO-C-H
H-C-OH
H-C-OH
CH_2OH
$\xrightarrow[\text{buffer,pH4}]{H_3PO_4/H_2O}$ $HOCH_2$-furan-CHO (O)

FORMATION OF ACETONIDE

Monosaccharides containing cis-glycol groups react with acetone in the presence of dehydrating agents to give isopropylidene derivatives called acetonides.

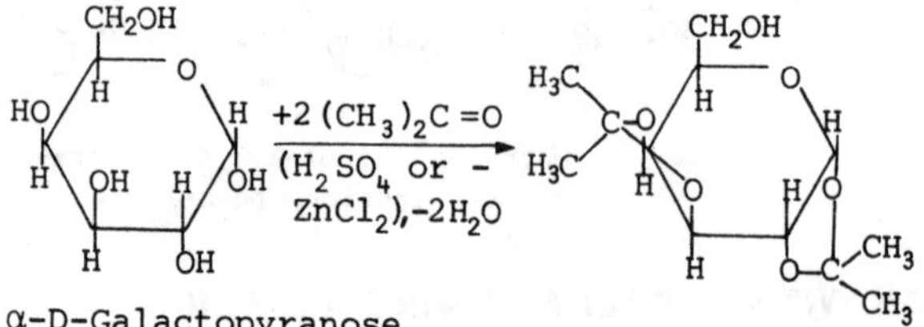

CH_2OH, O, H, HO, H, H, OH, H, OH, H, OH
$\xrightarrow[(H_2SO_4 \text{ or } - ZnCl_2), -2H_2O]{+2(CH_3)_2C=O}$
CH_2OH, O, H_3C, H_3C, C, O, H, H, H, O, H, O, H, O, C, CH_3, CH_3

α-D-Galactopyranose

1,2,3,4-Di-O-Isopropylidene-D-galactopyranose

OLIGOSACCHARIDES

Two monosaccharides condense to yield various types of disaccharides. If the hemiacetal -OH group of one molecule condenses with an alcoholic -OH of the second molecule, a maltose-type disaccharide is formed.

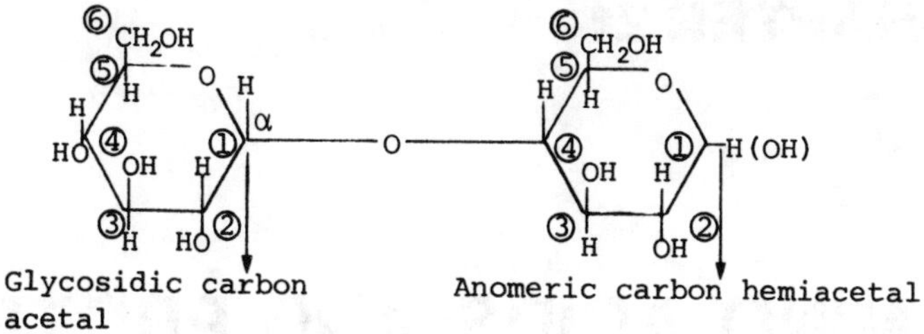
CH2OH
O
H
α
HO
OH
H
HO
O
CH2OH
O
H(OH)
OH
H
OH
Glycosidic carbon acetal
Anomeric carbon hemiacetal

POLYSACCHARIDES

Large numbers of monosaccharides joined by glycosidic linkages are called polysaccharides. They have large molecular weights.

Cellulose is a large linear unbranched natural polymer consisting of 3000-5000 D-glycose units per chain. Cellulose and lignin form the cell walls of wood and plants.

Starch is found in granular form in seeds and roots of plants. It has two components amylose (20%) and amylopectin (80%), both consisting of D-glucose units.

CHAPTER 21

AMINO ACIDS AND PROTEINS

21.1 AMINO ACIDS

Amino acids constitute a particularly important class of difunctional compounds because they are the 'building blocks' of proteins. The simpliest amino acid does not contain an asymmetric carbon and hence is optically inactive. All the other amino acids contain an asymmetric carbon, can exist as enantiomers and are optically active.

Table 21.1

THE COMMON AMINO ACIDS DERIVED FROM PROTEINS

Name	Abbreviation	Structure
Neutral Amino Acids		
Glycine	Gly	CH_2—COOH, with NH_2 on the CH_2
Alanine	Ala	CH_3—CH—COOH, with NH_2 on the CH
Valine†	Val	CH_3—CH—CH—COOH, with CH_3 on the first CH and NH_2 on the second CH
Leucine†	Leu	CH_3—CH—CH_2—CH—COOH, with CH_3 on the first CH and CH_2 on the second CH

Table 21.1 (continued)

Name	Abbreviation	Structure
Isoleucine†	Ileu	$CH_3-CH_2-CH(CH_3)-CH(NH_2)-COOH$
Phenylalanine†	Phe	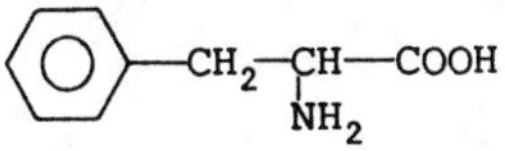 $C_6H_5-CH_2-CH(NH_2)-COOH$
Tyrosine	Tyr	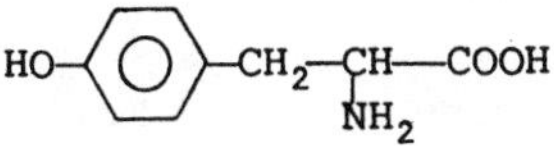 $HO-C_6H_4-CH_2-CH(NH_2)-COOH$
Tryptophan†	Try	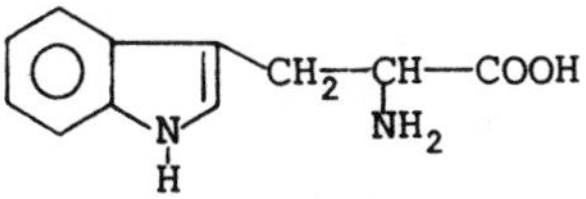 (indolyl, N–H)$-CH_2-CH(NH_2)-COOH$
Serine	Ser	$CH_2(OH)-CH(NH_2)-COOH$
Threonine†	Thr	$CH_3-CH(OH)-CH(NH_2)-COOH$
Proline	Pro	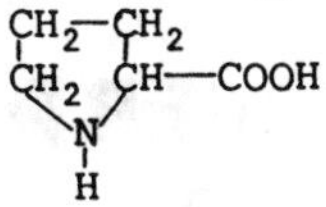 CH_2-CH_2 / CH_2 $CH-COOH$ / N–H (ring)
Neutral Amino Acids		
Hydroxyproline	HPro	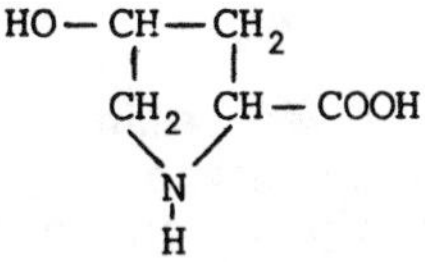 $HO-CH-CH_2$ / CH_2 $CH-COOH$ / N–H (ring)
Cysteine	CySH	$HS-CH_2-CH(NH_2)-COOH$
Cystine	CyS—SCy	$S-CH_2-CH(NH_2)-COOH$ / | / $S-CH_2-CH(NH_2)-COOH$
Methionine†	Meth	$CH_3-S-CH_2-CH_2-CH(NH_2)-COOH$
Acidic Amino Acids		
Aspartic acid	Asp	$HOOC-CH_2-CH(NH_2)-COOH$
Glutamic acid	Glu	$HOOC-CH_2-CH_2-CH(NH_2)-COOH$

Table 21.1 (continued)

Name	Abbreviation	Structure
Basic Amino Acid		
Lysine†	Lys	$CH_2-(CH_2)_3-CH(NH_2)-COOH$ with NH_2 on the first CH_2
Arginine	Arg	$HN{=}C(NH_2)-NH-(CH_2)_3-CH(NH_2)-COOH$
Histidine	His	$CH{=}C-CH_2-CH(NH_2)-COOH$ (imidazole ring: CH=C, N, NH, CH)

† Essential in human nutrition.

21.2 PROPERTIES OF AMINO ACIDS

$$R-\underset{\displaystyle NH_2}{\underset{|}{CH}}-COOH \qquad R-\underset{\displaystyle \oplus NH_3}{\underset{|}{CH}}-COO^-$$

The unionized amino acid is converted to a dipolar amino acid by an internal hydrogen ion transfer. Amino acids occur in the dipole ion form in aqueous solution or in the solid state.

Amino acids are generally high-melting solids because of the strong intermolecular attractions that can exist between the dipolar molecules of an amino acid. Amino acids decompose, rather than melt, at fairly high temperatures. They are more soluble in water than in non-polar organic solvents.

Amino acids in the dipolar ion form are amphoteric, which means they react with both acids and bases.

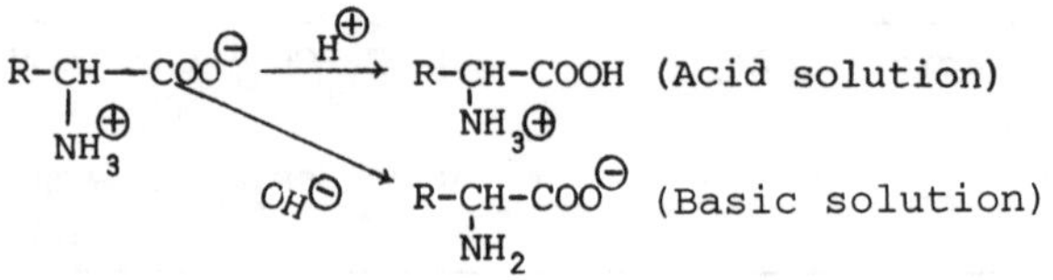

$$R-\underset{\displaystyle NH_3^{\oplus}}{\underset{|}{CH}}-COO^{\ominus} \xrightarrow{H^{\oplus}} R-\underset{\displaystyle NH_3^{\oplus}}{\underset{|}{CH}}-COOH \text{ (Acid solution)}$$

$$R-\underset{\displaystyle NH_3^{\oplus}}{\underset{|}{CH}}-COO^{\ominus} \xrightarrow{OH^{\ominus}} R-\underset{\displaystyle NH_2}{\underset{|}{CH}}-COO^{\ominus} \text{ (Basic solution)}$$

The isoelectric point of an amino acid is the hydrogen ion concentration of the solution in which the particular amino acid does not migrate under the influence of an electric field.

Neutral amino acids have isoelectric points at pH 5.5 to 6.3. Basic amino acids have isoelectric points at a high pH (around 10). Acidic amino acids have isoelectric points at a low pH (around 3).

All naturally-occurring amino acids belong to the L-series configuration.

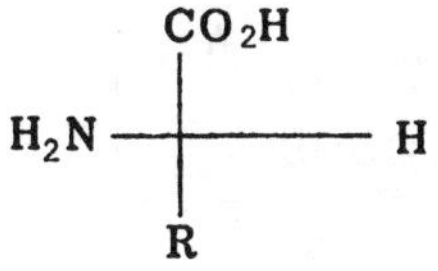

```
          CO2H                              CHO
           |                                 |
   H2N ----+------- H               HO ------+------- H
           |                                 |
           R                               CH2OH

   L-Amino acid                      L-Glyceraldehyde
```

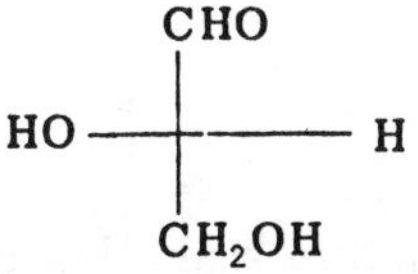

21.3 PREPARATION OF AMINO ACIDS

A) Ammonolysis of α-halo acids

$$CH_3CH_2COOH + Br_2 \xrightarrow{P} CH_3\underset{\displaystyle Br}{\underset{|}{C}}HCOOH + HBr$$

propionic acid — α-bromopropionic acid

$$\xrightarrow{NH_3(excess)} CH_3\underset{\displaystyle NH_2}{\underset{|}{C}}HCOOH + NH_4Br$$

alanine

B) Strecker synthesis

$$RCHO + NH_3 + HCN \rightarrow R\underset{\displaystyle NH_2}{\underset{|}{C}}HCN + HOH \xrightarrow{H_3O^+} R\underset{\displaystyle NH_2}{\underset{|}{C}}HCOOH$$

α-aminonitrile — α-amino acid

C) Acetylaminomalonate synthesis

$$EtOOCCHCOOEt(NHAc) \xrightarrow{NaOEt} \xrightarrow{EtBr} EtOOCC(Et)(NHAc)COOEt$$

$$\xrightarrow[heat]{H_3O^{\oplus}} EtCH(NH_2)COOH$$

The starting material can be made from malonic ester

$$CH_2(COOEt)_2 \xrightarrow{HONO} ON{-}CH(COOEt)_2 \xrightarrow{H_2, Ni} H_2NCH(COOEt)_2 \xrightarrow{Ac_2O} AcNHCH(COOEt)_2$$

D) Reductive amination of keto acids

Ex. $$CH_3COCOOH \xrightarrow{NH_3, H_2, Pt} CH_3CH(NH_2)COOH$$

pyruvic acid → alanine

E) The phthalimidomalonic ester method.

Potassium phthalimidate ($C_6H_4(CO)_2N^{\ominus}K^{\oplus}$) + $ClCH_2COOC_2H_5$ (Ethyl chloroacetate) → $C_6H_4(CO)_2NCH_2COOC_2H_5$

$$\xrightarrow{HCl, H_2O} Cl^{\ominus}\,{}^{\oplus}H_3NCH_2COOH$$

(glycine hydrochloride)
+ phthalic acid

21.4 REACTIONS OF AMINO ACIDS

A) Acid-base equilibria

$$CH_3CH(NH_2)COO^{\ominus} \underset{OH^-}{\overset{H^+}{\rightleftharpoons}} CH_3CH(NH_3^+)COO^- \underset{OH^-}{\overset{H^+}{\rightleftharpoons}} CH_3CH(NH_3^+)COOH$$

B) Acylation.

$$\overset{+}{H_3N}CH_2COO^- \xrightarrow{^{\ominus}OH} H_2NCH_2COO^- \xrightarrow{RCOCl} RCONHCH_2COOH$$

C) Esterification.

a) $$\overset{\oplus}{H_3N}CHRCOO^{\ominus} \xrightarrow[CH_3COCl]{NaOH} CH_3CONHCHRCOOH$$

$$\xrightarrow{SOCl_2} \xrightarrow{ROH} CH_3CONHCHRCOOR$$

b) $$H_3N^{\oplus}CH_2COO^- \xrightarrow{HCl} H_3\overset{\oplus}{N}CH_2COOH \xrightarrow[HCl]{CH_3OH} H_3\overset{\oplus}{N}CH_2COOCH_3$$

c) $$R{-}\underset{NH_2}{\underset{|}{C}}H{-}COOH + CH_3CH_2OH \xrightarrow{H^{\oplus}} R{-}\underset{NH_2}{\underset{|}{C}}H{-}COOCH_2CH_3 + H_2O$$

D) Reaction with nitrous acid (Van Slyke method).

$$R{-}\underset{NH_2}{\underset{|}{C}}H{-}COOH \xrightarrow[H_2O]{HNO_2} R{-}\underset{OH}{\underset{|}{C}}H{-}COOH + N_2\uparrow$$

E) Reaction with ninhydrin (Quantitative test).

Ninhydrin + R-CHCOOH (NH_2) α-Amino acid ⟶ Blue-to-violet color + RCHO + CO_2

F) Cyclic amides (or lactams)

a) α -amino acids form diketopiperazines.

Ex. $$2H_2NCH_2COOH \xrightarrow{heat} \text{(cyclo: } CH_2{-}CO{-}NH{-}CH_2{-}CO{-}NH\text{)} + 2HOH$$

b) β -amino acids lose ammonia to form unsaturated acids.

Ex. $H_2NCH_2CH_2COOH \rightarrow CH_2{=}CHCOOH + NH_3$

c) ß–amino acids form lactams;

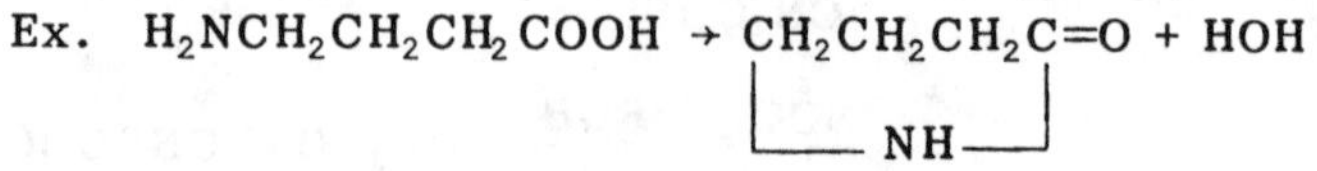

Ex. $H_2NCH_2CH_2CH_2COOH \rightarrow \underbrace{CH_2CH_2CH_2C{=}O}_{\text{—NH—}} + HOH$

γ–Butyrolactam

21.5 PEPTIDES

Peptides are amides formed by interaction between amino groups and carboxyl groups of amino acids. The amide group, $-\underset{\underset{O}{\|}}{C}-NH-$, is referred to as the peptide linkage.

A combination of two amino acids would give a dipeptide, three amino acids would give a tripeptide, and so on. A polypeptide is produced when a large number of amino acids are joined by peptide linkages.

The N-terminal amino acid residue that contains the free amino group is by convention, written at the left end. The C-terminal amino acid residue that contains the free carboxyl group is usually written at the right end.

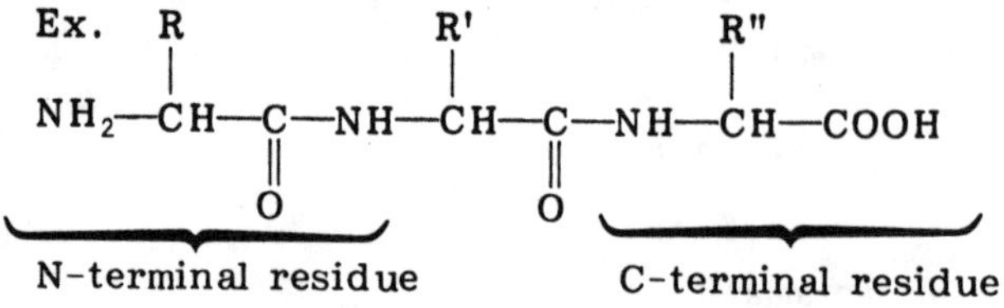

Ex.

$$\underbrace{NH_2-\underset{\overset{|}{R}}{CH}-\underset{\underset{O}{\|}}{C}}_{\text{N-terminal residue}}-NH-\overset{R'}{\overset{|}{CH}}-\underset{\underset{O}{\|}}{C}-\underbrace{NH-\overset{R''}{\overset{|}{CH}}-COOH}_{\text{C-terminal residue}}$$

A typical tripeptide

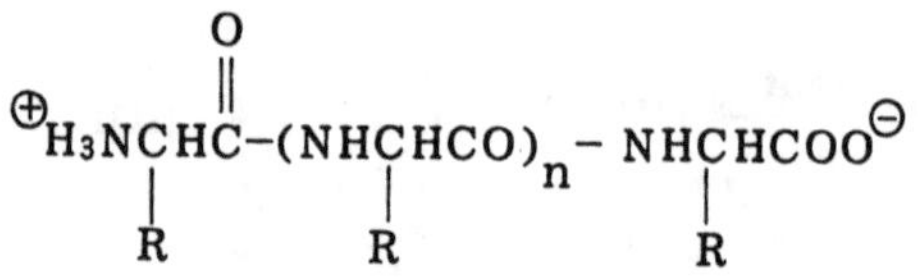

$$^{\oplus}H_3N\underset{\underset{R}{|}}{C}H\overset{\overset{O}{\|}}{C}-(NH\underset{\underset{R}{|}}{C}HCO)_n-NH\underset{\underset{R}{|}}{C}HCOO^{\ominus}$$

A polypeptide (n=1,2,3...)

When naming peptides, the amino acids present are listed starting with the N-terminal amino acid and going along the chain to the C-terminal amino acid. The amino acid suffix "-ine," is replaced by the suffix "-yl" for all the amino acids, except the C-terminal residue. For example,

$$\underbrace{NH_2-\underset{\substack{|\\CH_3}}{CH}-\overset{}{\underset{\substack{\|\\O}}{C}}}_{\text{Alanine}}-\underbrace{NH-CH_2-COOH}_{\text{Glycine}} \qquad \text{Alanylglycine}$$

$$\underbrace{NH_2CH_2-\underset{\substack{\|\\O}}{C}}_{\text{Glycine}}-\underbrace{NH-\overset{\substack{CH_2OH\\|}}{CH}-\underset{\substack{\|\\O}}{C}}_{\text{Serine}}-\underbrace{NH-\overset{\substack{CH_2SH\\|}}{CH}-COOH}_{\text{Cysteine}}$$

Glycylserylcysteine

The structural formula of a peptide can be represented by using the standard abbreviations of the constituent amino acids.

Ex.

Isoleucyllysylmethionyltyrosine (Ileu-Lys-Meth-Tyr)

Glutathione (glutamylcysteinylglycine) (Glu-CySH-Gly)

Glycylvalylcysteinylproline (Gly–Val–CySH–Pro)

21.6 STRUCTURE DETERMINATION OF PEPTIDES

The structural formula of peptides is determined by a combination of terminal residue analysis and partial hydrolysis.

Terminal residue analysis identifies the amino acid residues at the ends of the peptide chain. These procedures are used because the N- and C-terminal residues (which are at their respective ends of the peptide structure) differ from each other as well as from all the other residues.

A very successful method used to identify the N-terminal residue is pictured below:

$$O_2N-C_6H_3(NO_2)-F + H_2NCHCONHCHCO\sim \xrightarrow{\text{Alkaline medium}} O_2N-C_6H_3(NO_2)-NHCHCONHCHCO\sim$$

(R and R' side chains on the α-carbons)

2,4-Dinitro-fluorobenzene (DNFB) — N-Terminal residue, Peptide — "Labeled" peptide

Aq.HCl, heat

$$O_2N-C_6H_3(NO_2)-NHCH(R)COOH + {}^{\oplus}H_3NCH(R')COOH, \text{etc.}$$

N-(2,4-dinitrophenyl) amino acid (DNP, AA) — "Unlabeled" amino acids

The most widely used method of N-terminal residue analysis is as follows:

By this method, the rest of the peptide chain is left intact; then the analysis is repeated and the new terminal group of the shortened peptide is identified.

The C-terminal residue is determined enzymatically rather than chemically. The C–terminal residue of the peptide is selectively cleaved from the chain by using the enzyme carboxypeptidase as follows:

$$\text{Peptide chain}-NH-CH(R)-C(=O)-NH-CH(R')-COOH \xrightarrow[\text{enzyme}]{H_2O}$$

Carboxypeptidase-sensitive bond — C-Terminal residue

$$\text{Peptide chain}-NH-CH(R)-C(=O)-OH + NH_2-CH(R')-COOH$$

The structure of a peptide is determined by complete hydrolysis of the chain.

Hydrolysis of the DNP-peptide gives the DNP derivative of the N-terminal amino acid.

$$O_2N-C_6H_3(NO_2)-NH-Gly-Ser-CySH \xrightarrow{H_3O^{\oplus}}$$

DNP-Peptide

$$O_2N-C_6H_3(NO_2)-NH-Gly + Ser + CySH$$

DNP-Glycine

21.7 PREPARATION OF PEPTIDES

A) Chloroacyl chloride method.

Ex.

$$ClCH_2COCl + CH_3CH(NH_2)COOCH_3 \rightarrow ClCH_2CONHCH(CH_3)COOCH_3$$

$$\xrightarrow{NH_3} H_2NCH_2CONHCH(CH_3)COOCH_3$$

In the synthesis of peptides, a protecting group is employed. This 'protection' is required to insure that the desired reaction is not inadvertently carried out on the incorrect functional group of the amino acid.

B) Carbobenzoxy synthesis. The protection of an amino acid by the carbobenzoxy group.

$$C_6H_5CH_2OH + ClCOCl \rightarrow C_6H_5CH_2OCOCl + HCl$$

Phosgene (Carbonyl chloride) — Carbobenzoxy chloride (Benzyl chlorocarbonate)

The carbobenzoxy synthesis has been used to make glycylalanine as shown on the following page.

$$C_6H_5CH_2OCOCl + {}^{\oplus}H_3NCH_2COO^{\ominus} \rightarrow C_6H_5CH_2OCONHCH_2COOH$$

Carbobenzoxy chloride — Glycine — Carbobenzoxyglycine

$$\downarrow SOCl_2$$

$${}^{\oplus}H_3NCH(CH_3)COO^{\ominus} + C_6H_5CH_2OCONHCH_2COCl$$

Alanine — Acid chloride of carbobenzoxyglycine

$$\rightarrow C_6H_5CH_2OCONHCH_2CONHCH(CH_3)COOH \xrightarrow{H_2, Pd}$$

Carbobenzoxyglycylalanine

$${}^{\oplus}H_3NCH_2CONHCH(CH_3)COO^{\ominus}$$

Glycylalanine (Gly-Ala)

$$+ C_6H_5CH_3 + CO_2$$

C) The protection of an amino acid by the phthaloyl group.

$$\text{Phthalic anhydride} + NH_2-CH(R)-COOH \longrightarrow \text{Phthaloyl derivative } (C_6H_4(CO)_2N-CH(R)-COOH)$$

Phthalic anhydride — Phthaloyl derivative

$$C_6H_4(CO)_2N-CH(R)-C(=O)-NH-CH(R')-COOH + NH_2-NH_2 \longrightarrow$$

Phthaloyl-protected dipeptide — Hydrazine

$$NH_2-CH(R)-C(=O)-NH-CH(R')-COOH + C_6H_4(CO)_2(NH)_2$$

Dipeptide — phthalhydrazide

D) Solid-phase peptide synthesis (or the Merrifield method). The protection of an amino acid by a t-butoxycarbonyl group.

$$\text{t-BuOCONH}\underset{\text{R}}{\text{CH}}\text{COO}^{\ominus} + \text{ClCH}_2\text{C}_6\text{H}_4\text{-P} \longrightarrow \text{t-BuOCONH}\underset{\text{R}}{\text{CH}}\text{COOCH}_2\text{C}_6\text{H}_4\text{-P}$$

t-Butoxycarbonyl group (P=polymer)

$$\xrightarrow{\text{H}^{\oplus}} \text{H}_2\text{N}\underset{\text{R}}{\text{CH}}\text{COOCH}_2\text{C}_6\text{H}_4\text{-P} \xrightarrow[\text{dicyclohexylcarbodiimide (DCC)}]{\text{C}_6\text{H}_{11}\text{-N=C=N-C}_6\text{H}_{11} \; + \; \text{HOOC}\overset{\text{R'}}{\text{CH}}\text{NHOCO-t-Bu}}$$

$$\text{t-BuOCONH}\underset{\text{R'}}{\text{CH}}\text{CONH}\underset{\text{R}}{\text{CH}}\text{COOCH}_2\text{C}_6\text{H}_4\text{-P} \xrightarrow{\text{H}^{\oplus}}$$

$$\underbrace{\xrightarrow{\text{H}^{\oplus}} \; \xrightarrow{\text{DCC} + \text{HOOC}\overset{\text{R''}}{\text{CH}}\text{NHOCO-t-Bu}}}_{\text{To be repeated many times}} \; \xrightarrow[\text{Final step}]{\text{HBr}}$$

$$\underbrace{\text{H}_2\text{N-}\underset{\text{R}^{\text{L}}}{\text{CH}}\text{CO}}_{\text{Last residue to be added}}\sim\sim\text{NH}\underset{\text{R''}}{\text{CH}}\text{CONH}\underset{\text{R'}}{\text{CH}}\text{CONH}\underset{\text{R}}{\text{CH}}\text{COOH} + \text{BrCH}_2\text{C}_6\text{H}_4\text{-P}$$

21.8 PROTEINS

Proteins are polypeptides of great size and weight. They differ from other polypeptides by having higher molecular weights (over 10,000) and more complex structures. Some of the chemical behavior of proteins is similar to that of polypeptides. For example, the techniques of hydrolysis and end-group analysis can be applied to proteins.

Many proteins undergo denaturation (or precipitation), which is a nonchemical transformation that may alter the physical or biological properties of a protein. Few, if any, changes in the chemical structure result from this process. Protein denaturation can be caused by heat, ultraviolet light, organic solvents, extremes of pH and/or by many other chemical agents. Denaturation is either reversible or irreversible, depending upon the agent used. The coagulation of an egg white by heating is an example of irreversible denaturation.

Polypeptides do not undergo denaturation, probably due to their smaller and less complex nature.

Pure proteins are usually amorphous solids. Proteins are optically active and divided into two basic classes: fibrous proteins, which are insoluble in water and are resistant to denaturation; and globular proteins, which are soluble in water and in aqueous solutions of acids, bases, or salts. Globular proteins are more sensitive to denaturation than are fibrous proteins.

When proteins are dissolved in water, they form colloidal solutions due to their large molecular size.

Molecules of fibrous proteins are long and thread-like and tend to lie side by side to form fibers; in some cases they are held together by hydrogen bonds. The intermolecular forces are, therefore, very strong.

Fibrous proteins function as the main structural materials of animal tissues because of their insolubility and fiber-forming tendency.

Globular proteins serve functions that require mobility and solubility. These functions are related to the maintenance and regulation of the life processes. Such proteins can be enzymes.

Proteins, like amino acids, have isoelectric points. These are the specific pH's at which the acid functional groups neutralize the amino functional groups. As with amino acids, solubility is at minimum at the isoelectric point. Proteins, like amino acids, are amphoteric, this is due to the fact that they contain both free carboxyl (acidic) and free amino (basic) groups.

21.9 STRUCTURE OF PROTEINS

PRIMARY STRUCTURE OF PROTEINS

The primary structure of a protein is the amino acid sequence of the protein, which is genetically determined for every protein.

The primary structure of a protein may contain more than one amino acid chain. These chains are bonded to each other at specific points by disulfide, -S-S-, linkages. The following diagram shows how two amino acid chains can be joined by these linkages.

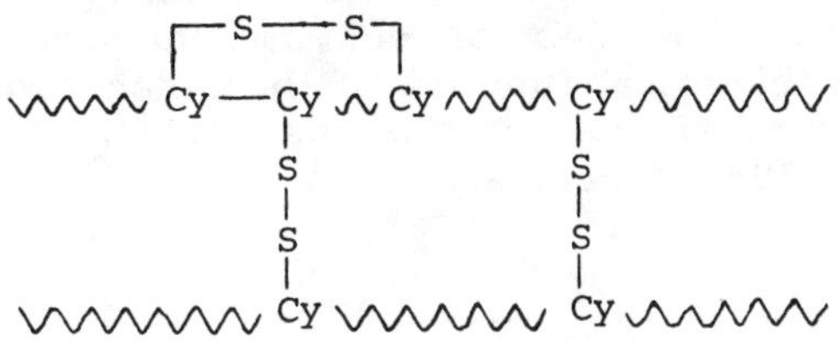

Example of disulfide linkages in a protein.

The amino acid sequence for any given protein is always the same, and any slight change in its sequence may have dramatic biological consequences. A change of 1 amino acid out of 150, completely alters the properties of the protein.

Ex. Val-His-Leu-Thr-Pro-(Glu)-Glu-Lys...

Normal hemoglobin

Val-His-Leu-Thr-Pro-(Val)-Glu-Lys...

Sickle-cell hemoglobin

Sickle-cell anemia is the hereditary disease caused by the substitution of a valine for a glutamic acid in hemoglobin.

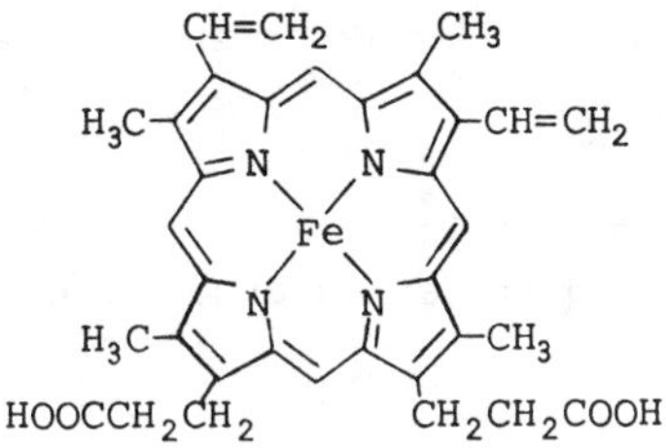

Heme

A prosthetic group is the non-peptide part of a conjugated protein molecule. The prosthetic group is concerned with the biological action of the protein. An example of a prosthetic group is the heme in hemoglobin.

Many enzymes require co-factors to exert their catalytic effects. If the organic co-factors are covalently bonded to the enzyme, they are called coenzymes; these too are prosthetic groups.

The coenzyme nicotinamide adenine dinucleotide (NAD) is associated with a number of dehydrogenation enzymes. The chracteristic biological function of these dehydrogenation enzymes is to convert the nicotinamide portion of NAD or NADP into the dihydro structure.

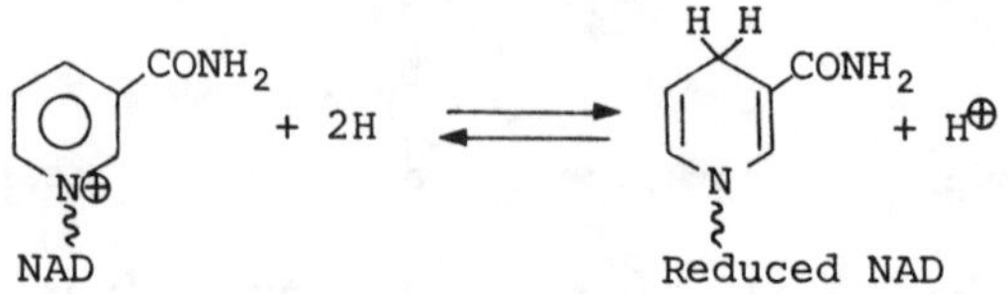

$$\text{NAD} + 2H \rightleftharpoons \text{Reduced NAD} + H^{\oplus}$$

Many of the molecules which make up coenzymes are vitamins.

SECONDARY STRUCTURE OF PROTEINS

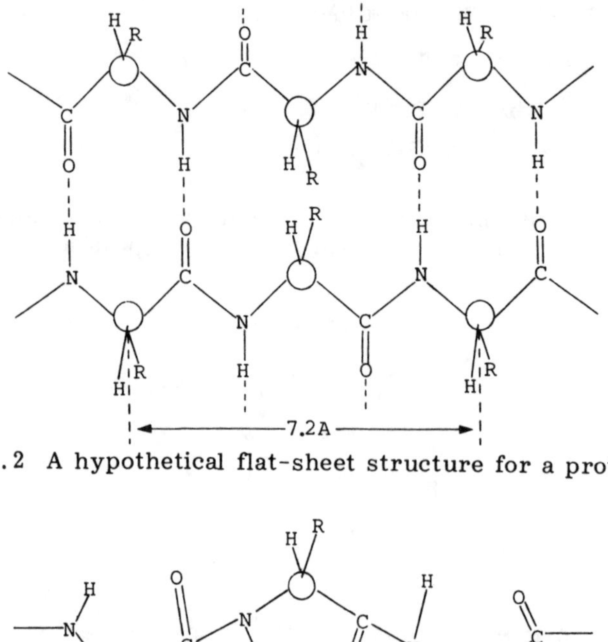

7.2A

Fig. 21.2 A hypothetical flat-sheet structure for a protein.

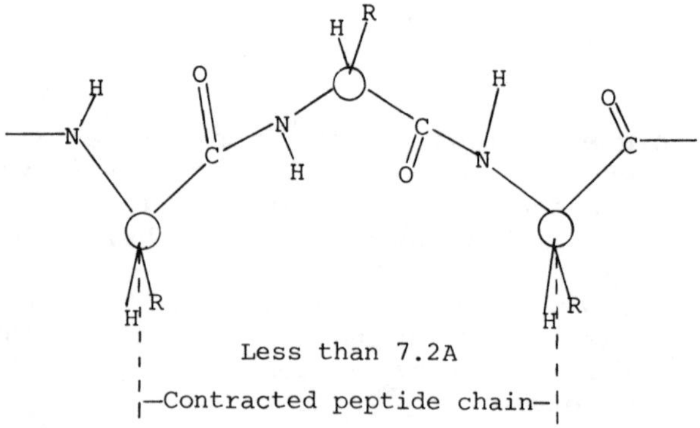

Less than 7.2A

Contracted peptide chain

Fig. 21.3 This structure is called the beta-arrangement, it is characterized by a pleated-sheet form.

Many amino acid chains exist in the form of a right-handed helical coil which is stabilized by hydrogen bonds. The coil is called the alpha-helix and contains 3.6 amino acid residues per turn of the helix.

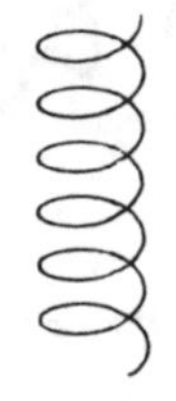

∝ helix
(right-handed)

In addition to the x-ray diffraction patterns characteristic of the alpha- and beta-type proteins, there is a third kind called collagen, the protein of tendon and skin. Collagen is characterized by a high proportion of proline and hydroxy proline residues, and by frequent repetitions of the sequence Gly-Pro-Hypro. The pryrrolidine ring of proline and hydroxyproline can affect the secondary structure.

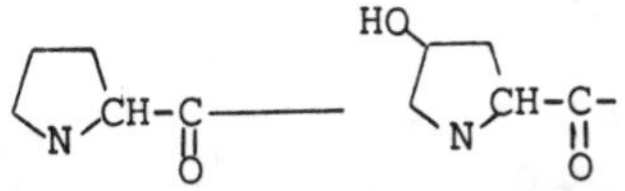

Proline residue   Hydroxyproline residue

TERTIARY STRUCTURE OF PROTEINS

The tertiary structure is the manner in which the protein helical coils are arranged to give a gross protein structure. The protein is globular if the helical coil is intertwined into a sphere. Fibrous proteins are formed by the winding together of helical coils to form long strands.

Globular
Protein

Fibrous protein

The mechanism of denaturation involves the disruption of the spherical shape (tertiary structure).

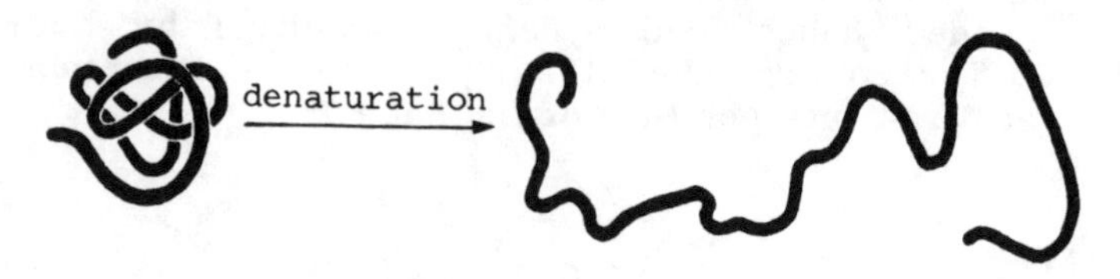

QUATERNARY STRUCTURE OF PROTEIN

The quaternary structure is the clumping together of the spherical units of globular proteins into specific shapes. The cause of the clustering is not as yet clearly understood, however, it may possibly be due to electrostatic attraction.

The following are examples of the quaternary structure of proteins:

1-

Myoglobin, which has no quarternary structure.

2-

Hemoglobin

3-

Polio virus

4-

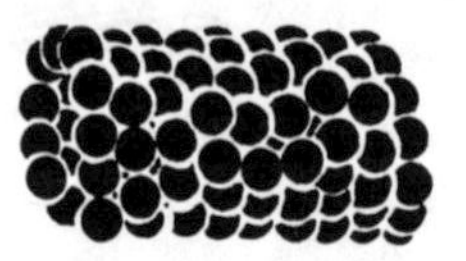

Tobacco-mosaic virus.

CHAPTER 22

SPECTROSCOPY

22.1 MASS SPECTRUM

Each kind of ion has a particular ratio of mass to charge, or m/e value. The m/e value is simply the mass of the ion.

Molecular ion

	$(C_4H_9)^{\oplus}$	$(C_3H_5)^{\oplus}$	$(C_2H_5)^{\oplus}$	$(C_2H_3)^{\oplus}$	and others
m/e:	57	41	29	27	
Relative intensity:	100	41.5	38.5	15.7	
	Base peak				

The peak of the largest intensity is called the base peak; its intensity is taken as 100. The intensities of the other peaks are expressed relative to the base peak. A mass spectrum is a plot; the abscissa is the m/e ratio and the ordinate is the relative number of ions of relative intensity (height of each peak).

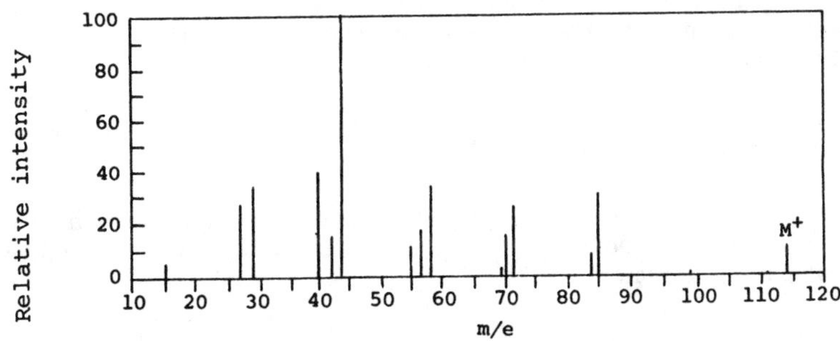

Fig. 22.1 The mass spectrum of n-octane.

The mass spectrum is used to determine the molecular weight, molecular formula and structure of a compound.

When a molecule loses an electron, a molecular ion is produced. The molecular ion peak is called the parent peak and its m/e value is the molecular weight of the compound. The existence of isotopes produce small peaks with m/e values of M + 1, M + 2, etc.

$$M + e^{\ominus} \rightarrow M^{\oplus} + 2e^{\ominus}$$

Molecular ion
(Parent ion)

m/e = mol. wt.

22.2 ELECTROMAGNETIC SPECTRUM

Energy is associated with the electromagnetic radiation at a particular frequency by the following expression:

$$\Delta E = hv,$$

where ΔE = gain in energy, ergs.
h = Plank's constant, 6.5×10^{-27} ergs-sec
v = frequency, Hz (cycles/sec)

Or by

$$\Delta E = \frac{hc}{\lambda},$$

where c = speed of light, 3×10^{10} cm/sec

λ = wavelength, cm.

The higher the frequency (the shorter the wavelength), the greater the gain in energy.

The structure of a molecule determines the electronic, vibrational, and rotational levels in the molecule.

The electromagnetic spectrum of a compound shows how much electromagnetic radiation is absorbed at each frequency. It determines the compound's structure.

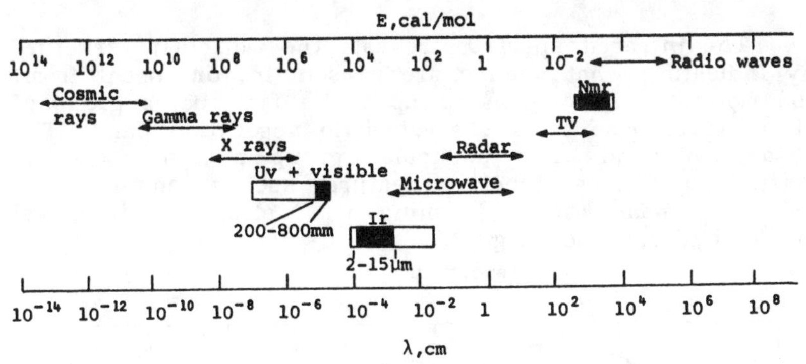

Fig. 22.2 Electromagnetic spectrum.

22.3 INFRARED SPECTRUM

The infrared spectrum of a compound is produced by the infrared absorption of its molecules, due to the degree of molecular vibration and rotation of its bonds when subjected to infrared irradiation.

The IR spectrometer records the percent transmittance (%T) of incident light through the sample as a function of the wavelength of light, expressed in micrometers ($\mu m = 10^{-4}$cm), or versus the frequency of the incident light, expressed in wave numbers. Transmittance is related to absorbance, A, by the equation $A = \log_{10} 1/T$. The wave number is the reciprocal of the wavelength in centimeters and is directly proportional to the energy of the light absorbed.

Table 22.1 Infrared Absorptions of Diagnostic Value

Functional Group	Infrared Absorption, μ
O—H	2.8-3.1
N—H	2.9-3.2
C—H	3.0-3.5
C ≡ C	4.5
C ≡ N	4.5
C=O	5.7-6.0
C=C (alkenes)	6.0-6.2
C⎓C (aromatic)	6.2-6.3 and 6.7-6.8
$-\overset{\oplus}{N}(=O)O^{\ominus}$	6.5 and 7.5

The infrared spectrum reveals the molecular structure by indicating what groups are present in, or absent from, the molecule. This is based upon the fact that a group of atoms gives rise to a characteristic absorption band. The absorption band of a particular group of atoms can be shifted by various structural features such as angle strain, van der Waals strain, conjugation, electron withdrawal and/or hydrogen bonding.

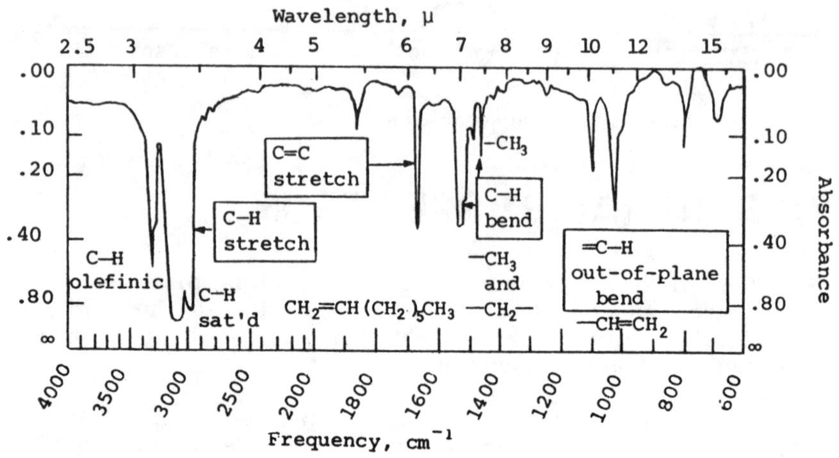

Fig. 22.3 The infrared spectrum of 1-octene.

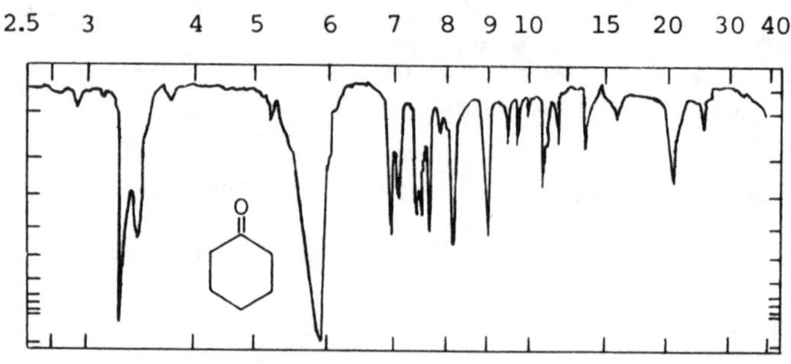

Fig. 22.4 The infrared spectrum of cyclohexanone.

22.4 ULTRAVIOLET SPECTRUM

Ultraviolet spectroscopy involves the excitation of an electron in its ground state level to a higher energy level. This is accomplished by irradiating a sample with ultraviolet light (electromagnetic radiation with wavelengths in the range of 200nm to 400nm).

A spectrum is described by the position of the maximum absorption (λ_{max}) and the intensity or the molar absorptivity of the incident light (ε_{max}, the extinction coefficient).

The molar absorptivity is related to absorbance by Beer's law.

$$\varepsilon = \frac{A}{Cl},$$

where ε = molar absorptivity (molar extinction coefficient)

A = absorbance at λ_{max}

C = molar concentration of sample, $\frac{\text{moles}}{\text{liter}}$

l = path length of sample tube, cm.

$$A = \log(I_0/I),$$

where I_0 = initial light intensity.

I = final light intensity.

Therefore,

$$\varepsilon = \frac{\log(I_0/I)}{Cl}$$

The percentage of light absorbed is related to the percentage of light transmitted by the following equation:

$$\%A = 100 - \%T,$$

where $\%A$ = % absorbance

$\%T$ = % transmittance

Also,

$$T = \frac{I}{I_0}$$

Molecules contain bonding electrons which are directly involved in their bonding and which are in sigma (σ) or pi (π) molecular orbitals. σ^* and π^* are antibonding orbitals and are unstable where the electron density between the nuclei is very low. Many molecules that contain the O, S, N, Br, Cl, F and I atoms, contain non-bonding electrons which are not directly involved in bonding and which are in unhybridized non-bonding orbitals, n.

The n → π* and π → π* transitions are the most observed and useful transitions in organic molecules. A chromophore is the molecular or functional group that gives rise to π → π* and/or n → π* transitions. C=C, C≡C, C=O, N=O, C=S, and aromatic rings are typical chromophores.

In the n → π* translation, the electron of an unshared pair goes to an unstable (antibonding) π* orbital.

$$C{=}\ddot{O}: \rightarrow C{\dot{=}}\dot{O}: \qquad n \rightarrow \pi^*$$

In the π → π* transition, an electron goes from a stable (bonding) π orbital to an instable π* orbital.

$$>C{=}\ddot{O}: \rightarrow C\overset{:}{—}\ddot{O}: \qquad \pi \rightarrow \pi^*$$

Conjugation of double bonds lowers the energy required for the transition, and absorption moves to longer wavelengths.

Resonance stabilizes the excited state more than the ground state and it reduces the difference between them.

The ultraviolet spectrum shows the relationships (mainly conjugation) between functional groups. It reveals the number and location of substituents attached to the carbons of conjugated systems.

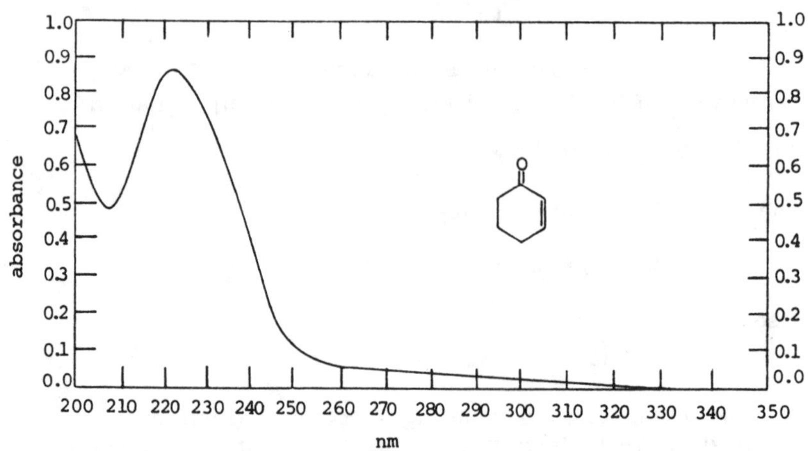

Fig. 22.5 The ultraviolet spectrum of cyclohexanone.

Very highly conjugated molecules absorb light in the visible region, 400 to 800nm. When absorption occurs in the visible region, the compound appears to be colored. The color of a compound corresponds to the wavelength of the light that remains after the absorbed light has been subtracted. The compound has a complementary color to the color of the light absorbed.

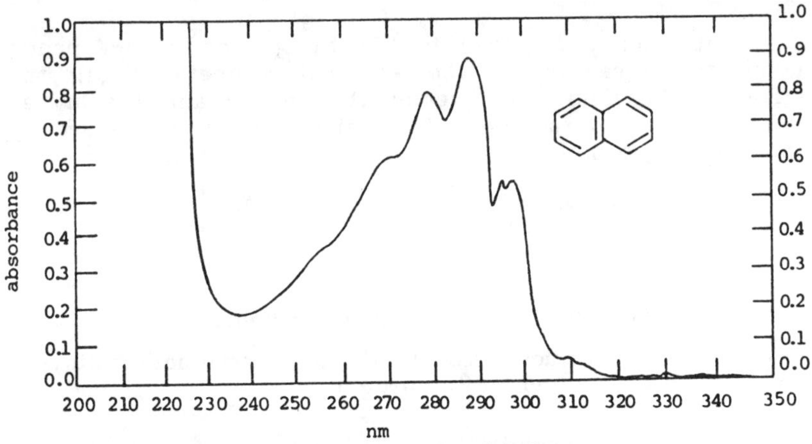

Fig. 22.6 The ultraviolet spectrum of naphthalene.

Table 22.2 Wavelengths of Visible Colors

Wavelength, nm	Color of the light	Complementary color
400	Violet	Yellow
450	Blue	Orange
500	Blue-green	Red
550	Yellow	Violet
600	Orange-red	Blue-green
700	Red	Green

22.5 NUCLEAR MAGNETIC RESONANCE (NMR) SPECTRUM

When an atom is placed in an external magnetic field, the magnetic field generated by the nucleus will be aligned with or against the external magnetic field. Alignment with the field is more stable. At some frequency of electromagnetic radiation, the nucleus will absorb energy and "flip" over so that it reverses its alignment with the external field. This is known as the nuclear magnetic resonance (nmr) phenomenon. It is concerned with the nuclear magnetic resonance of hydrogen atoms.

The energy required to flip the proton over depends upon the strength of the external magnetic field; the stronger the field, the greater the energy and the higher the frequency of radiation. This relationship is expressed by the following equations:

$$\nu = \frac{\lambda H_0}{2\pi}$$

where ν = frequency, Hz.

H_0 = strength of the magnetic field, gauss.

λ = a nuclear constant, the gyromagnetic ratio, 26,750 for the proton.

$$\nu = \frac{2\mu H_0}{h}$$

where μ = the magnetic moment of the nucleus.

h = Planck's constant.

In nuclear magnetic resonance, spectroscopy molecules are placed in a strong magnetic field to create different energy states which are then detected by absorption of light of the appropriate energy.

Various aspects of the nmr spectrum are:

A) the number of signals, which tell us how many different "kinds" of protons are present in a molecule.

B) the positions of the signals, which tell us something about the electronic environment of each kind of proton;

C) the intensities of the signals, which tell us how many protons of each kind are present; and

D) the splitting of a signal into several peaks, which tells us about the environment of a proton with respect to other, nearby protons.

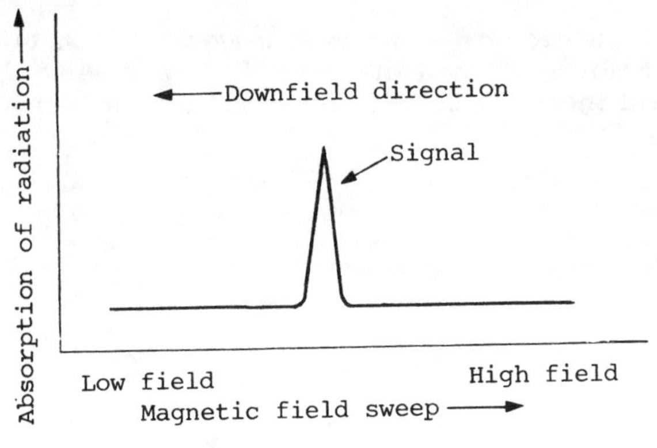

Fig. 22.7 The nmr spectrum.

NUMBER OF SIGNALS

A set of protons with the same environment are equivalent.

$CH_3-CH_2-CH_2-Cl$ (a b c)	$CH_3-CHCl-CH_3$ (a b a)
3 nmr signals	2 nmr signals
n-Propyl chloride	Isopropyl chloride

In order for protons to be chemically equivalent, they must be stereochemically equivalent.

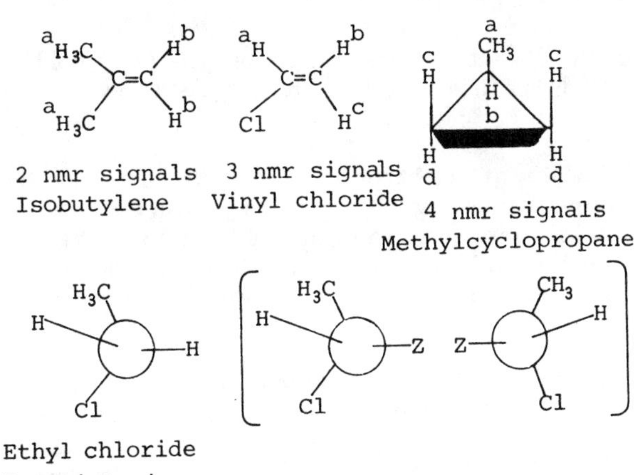

These two protons are mirror images of each other, in an achiral medium. They behave as if they were equivalent, and there will therefore be one nmr signal for the pair.

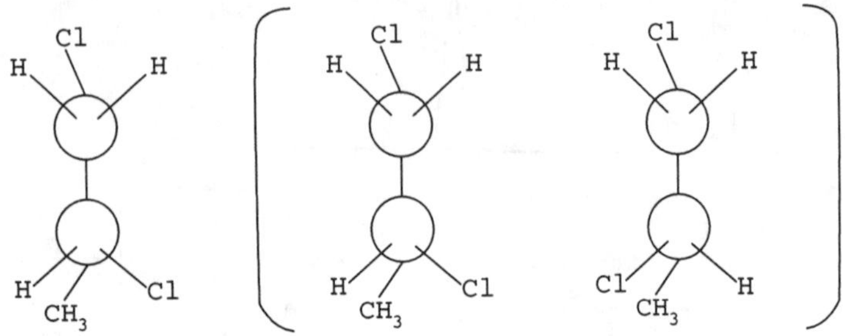

1,2-Dichloropropane
Diastereotopic
protons

These two protons are neither identical nor mirror images of each other. They are non-equivalent, and there will be a separate nmr signal for each one.

POSITIONS OF SIGNALS

The positions of signals determine what kind of protons they represent: aromatic, aliphatic, primary, secondary, tertiary, benzylic, vinylic, acetylenic, or adjacent to halogen or other atoms or groups.

Circulation of electrons about the proton generates a field that opposes the applied field. The field felt by the proton is decreased, and the proton is shielded.

Circulation of electrons - specifically, π electrons - about nearby nuclei generates an induced field. If this induced field opposes the applied field, the proton is shielded. If the induced field strengthens the applied field, then the effective field of the proton is increased, and the proton is deshielded.

A shielded proton requires a higher applied field strength, and a deshielded proton requires a lower applied field strength. Shielding shifts absorption upfield, and deshielding shifts absorption downfield. Chemical shifts are shifts in the position of nmr absorptions that arise from shielding and deshielding by electrons.

The point from which chemical shifts are measured is the reference point.

The δ (delta) scale is the most commonly used scale and is calculated by the equation

$$\delta = \frac{(\text{chemical shift in Hz}) \times 10^6}{\text{spectrometer frequency}}$$

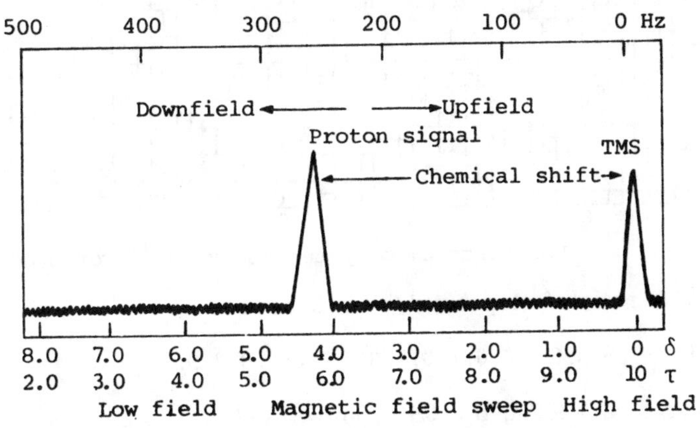

Fig. 22.8 An example of the nmr spectrum.

POSITIONS OF ABSORPTION OF HYDROGEN ATOMS IN NUCLEAR-MAGNETIC-RESONANCE SPECTROSCOPY

Type of Hydrogen Atom	Absorption, in ppm from TMS
$(CH_3)_4Si$	0
$CH_3-\overset{\vert}{\underset{\vert}{C}}-$	0.6-1.5
$C-CH_2-C$	1.2-1.5
$C-\overset{C}{\underset{H}{C}}-C$	1.4-1.8
$CH_3-\underset{\Vert \atop O}{C}-$	1.9-2.5
$H-C\equiv C-$	2.5-3.0
$C=C\begin{smallmatrix}H\\H\end{smallmatrix}$	4.5-6.6
C_6H_5-H	6.5-8.0

PEAK AREA AND PROTON COUNTING

The area under an nmr signal is directly proportional to the number of protons that give rise to the signal.

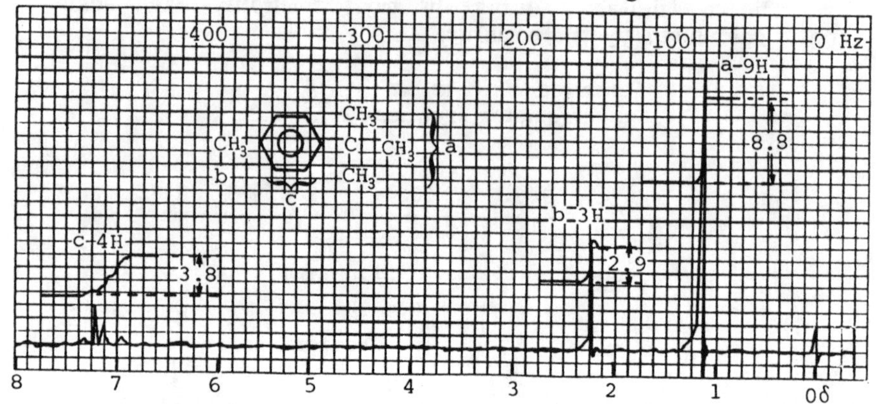

Fig. 22.9 Nmr spectrum of p-tert-butylene. Proton counting.

The ratio of step heights, a:b:c, is

8.8:2.9:3.8 = 3.0:1.0:1.3 = 9.0:3.0:3.9.

Therefore, a = 9H's, b = 3H's, and c = 4H's.

Splitting of signals. Spin-spin coupling.

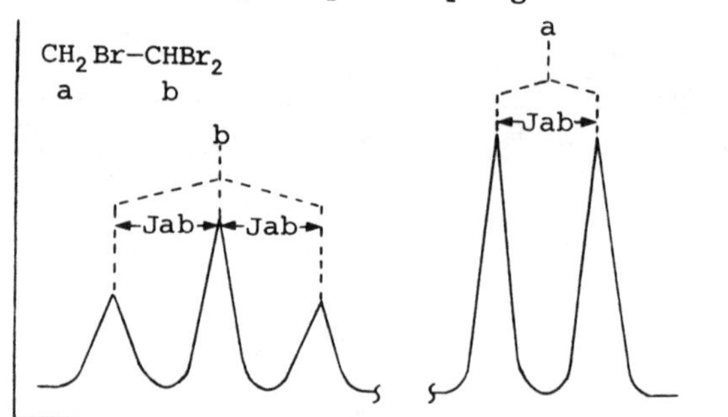

Fig. 22.10 Nmr splitting of signals. Spin-spin splitting.

An nmr signal is split into a doublet by one nearby proton, and into a triplet by two (equivalent) nearby protons.

$CH_3C(=O)CH_3$ → Singlet (6H)

CH_3CH_2I → Triplet (3H), Quartet (2H)

$ClCH_2CH_2CH_2Cl$ → Quintet (2H) (middle CH_2); Triplet (4H) (terminal CH_2's)

The number of peaks (or multiplicity) observed for a given proton (or equivalent protons) is equal to n + 1, where n is the number of protons on adjacent atoms.

Spin-spin splitting is observed only between non-equivalent neighboring protons. Non-equivalent protons are protons with different chemical shifts. Neighboring protons are protons on adjacent carbons.

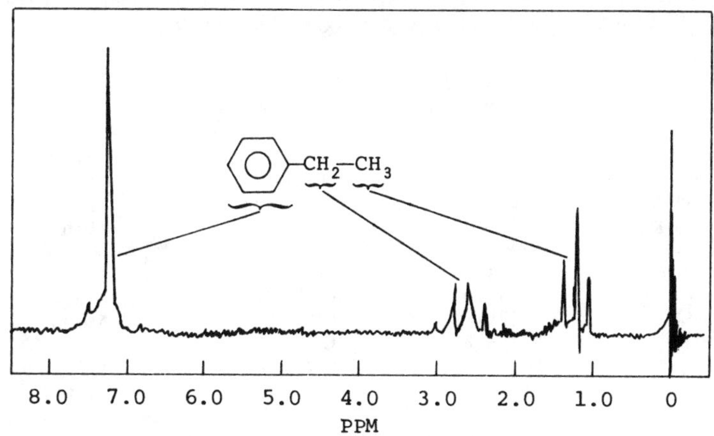

Fig. 22.11 The nmr spectrum of ethyl benzene.

COUPLING CONSTANT

The distance between peaks in a multiplet is called the spin coupling constant, J. It is a measure of the effectiveness of spin-spin coupling.

Peak separations that result from splitting remain constant, while peak separations that result from chemical shifts change.

The size of a coupling constant depends on the structural relationships between coupled protons. The size of J varies with the electronegativity of substituents.

CARBON-13 NMR

Carbon-13 NMR has several advantages over proton NMR in determining the structure of organic molecules (including biological compounds). Perhaps the most important advantage is that carbon-13 NMR provides a great deal more information about the structure of a molecule than does proton NMR which only provides information on peripheral protons. This occurs since the backbone of organic molecules is composed of carbon.

Another advantage is that the chemical shift in most organic compounds is about 200 ppm (compare this with 10 to 20 ppm for proton NMR), so there is less chance that peaks will overlap. Due to the low abundance of carbon-13 in natural samples (1%), it is unlikely that two carbon-13 atoms will be next to each other on the same molecule, hence homonuclear spin-spin coupling is not encountered. Since carbon-12 has zero quantum spin, heteronuclear spin-spin coupling is not a problem with this isotope. Also, good methods exist for decoupling the interactions between carbon-13 and protons.

type of carbon	*chemical shift in ppm*	*type of carbon*	*chemical shift in ppm*
CH_3–halogen	0 to 40	–C≡C–	75 to 95
CH_3N=	10 to 45	–C≡N	115 to 125
–CH_2–halogen	0 to 50	C=C alkene	110 to 145
=CH–halogen	35 to 75	C=C aromatic	110 to 160
–CH_2–N=	40 to 50	≡C–C≡	30 to 50
=CH–N=	60 to 80	=CH–C≡	30 to 45
CH_3–O	50 to 65	–CH_2–C≡	20 to 45
–CH_2–O	40 to 75	CH_3–C≡	5 to 30
=CH–O	75 to 85	C=O acid	170 to 180
		C=O aldehyde	195 to 205
		C=O ketone	195 to 215

FOURIER TRANSFORM (FT) NMR

Carbon-13 NMR would never have been practical (due to the low abundance of carbon-13) had it not been for the advancement of FT NMR. In FT NMR, resolution of the elements in the spectrum is the result of very brief measurements that yield a time domain spectrum, rather than a frequency domain. This time domain spectrum can be obtained in less than a few seconds, hence many (hundreds, or even thousands) can be taken and averaged in a relatively brief period of time. A frequency domain spectrum can then be obtained by a Fourier transform using a digital computer.

22.6 ELECTRON SPIN RESONANCE (ESR) SPECTRUM

The odd electron of a free radical generates a magnetic moment by spinning. Each electron of an electron pair generates a magnetic moment of equal and opposite magnitude (because the two electrons have equal and opposite spins). When a free radical is placed in a magnetic field, the magnetic moment generated by the odd electron may be aligned with or against the external magnetic field. When this system is exposed to electromagnetic radiation of the proper frequency, the odd electron absorbs radiation and reverses its spin; like the proton in nmr, the electron "flips" over in esr. An absorption spectrum is obtained and is called an electron spin resonance (esr) spectrum or an electron paramagnetic resonance (epr) spectrum.

The signals of an esr spectrum show splitting for the same reason that nmr signals split. The esr signal will be split by n neighboring protons into n + 1 peaks.

Esr spectroscopy is used to detect the presence of free radicals, to measure their concentration, and to determine their structure. The protons that are responsible for splitting the signal indicate the distribution of the odd electron in the free radical.

The eight line esr signal of propane indicates that there are seven neighboring hydrogens about the odd electron.

$CH_3\dot{C}HCH_3$

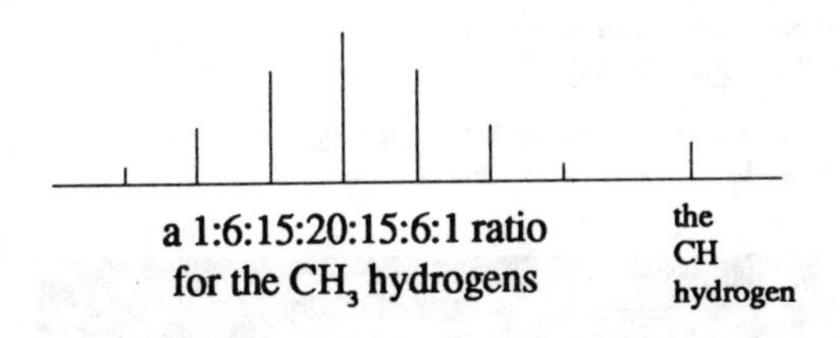